The Urbana Free Library

To renew: call **217-367-4057**
or go to **urbanafreelibrary.org**
and select **My Account**

ADVENTURES IN THE RESPIRATORY SYSTEM

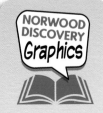

NORWOOD DISCOVERY Graphics

THE BOD SQUAD

by Alexander Lowe • illustrated by Sebastian Kadlecik

Norwood House Press

For more information about Norwood House Press please visit our website at: www.norwoodhousepress.com or call 866-565-2900.

Library of Congress Cataloging-in-Publication Data
Names: Lowe, Alexander, author. | Kadlecik, Sebastian, illustrator.
Title: Adventures in the respiratory system / Alexander Lowe ; illustrated by Sebastian Kadlecik.
Description: Chicago : Norwood House Press, 2020. | Series: Norwood discovery graphics | Audience: Ages 8-10 | Audience: Grades 4-6
 | Summary: "When Sam breathes in a speck of dust through his nose, The Bod Squad shrinks down in size to see how his respiratory
 system reacts. The squad learns how the body separates out the dust from the oxygen it needs and how oxygen travels through the
 body. An adventure-filled graphic novel that provides information about the human body and how its respiratory system works.
 Includes contemporary full-color graphic artwork, fun facts, additional information, and a glossary"— Provided by publisher.
Identifiers: LCCN 2020024500 (print) | LCCN 2020024501 (ebook) | ISBN 9781684508600 (hardcover)
 | ISBN 9781684045822 (paperback) | ISBN 9781684045877 (epub)
Subjects: LCSH: Respiratory organs—Comic books, strips, etc. | Respiratory organs—Juvenile
 literature. | Lungs—Physiology—Juvenile literature. | Graphic novels.
Classification: LCC QP121 .L78 2020 (print) | LCC QP121 (ebook) | DDC 612.2—dc23
LC record available at https://lccn.loc.gov/2020024500
LC ebook record available at https://lccn.loc.gov/2020024501

Hardcover ISBN: 978-1-68450-860-0   Paperback ISBN: 978-1-68404-582-2

328N—072020
Manufactured in the United States of America in North Mankato, Minnesota.

CONTENTS

MEET THE BOD SQUAD

Jada

Kara

Logan

Sam

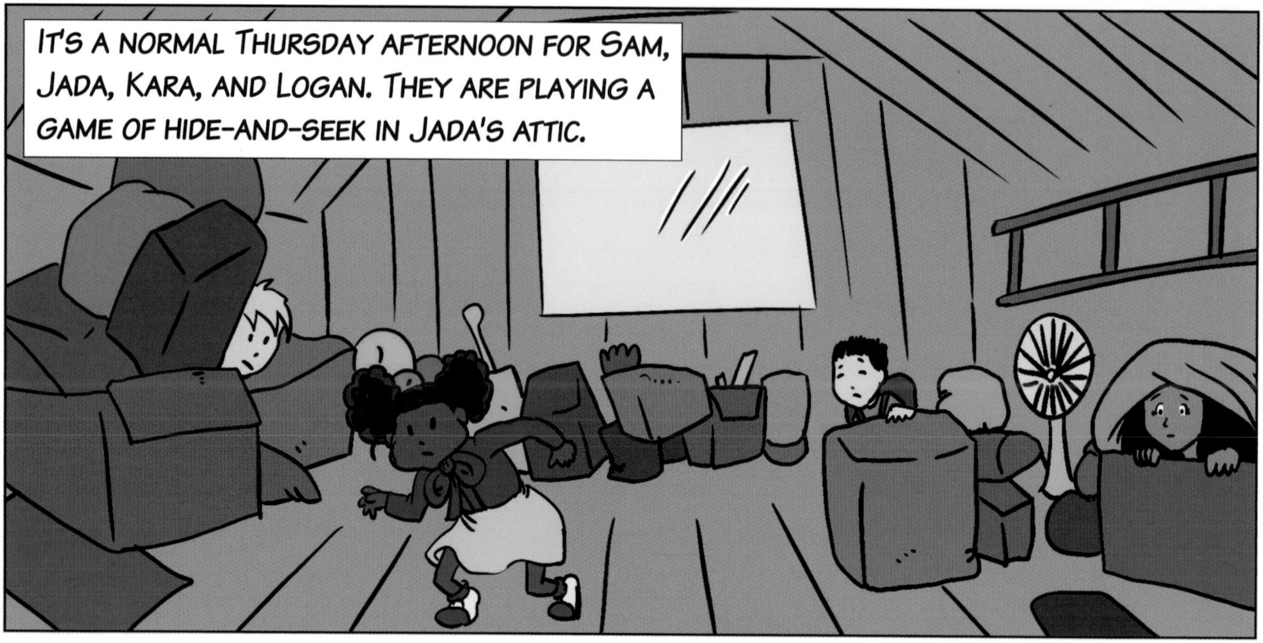
IT'S A NORMAL THURSDAY AFTERNOON FOR SAM, JADA, KARA, AND LOGAN. THEY ARE PLAYING A GAME OF HIDE-AND-SEEK IN JADA'S ATTIC.

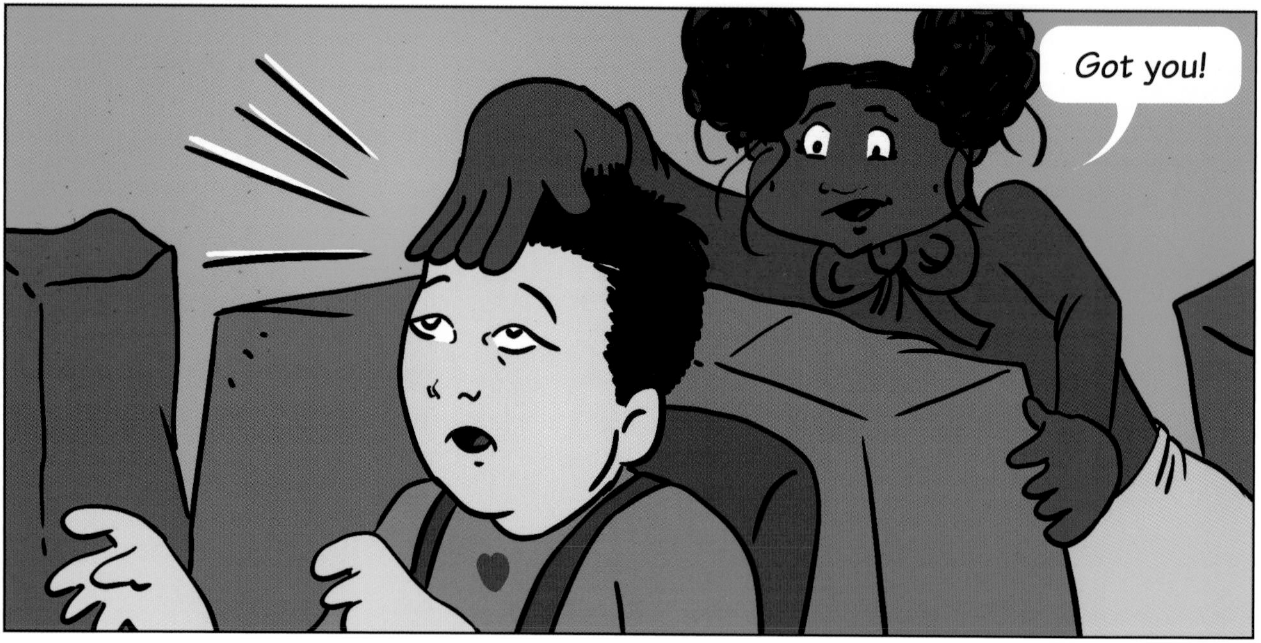
Got you!

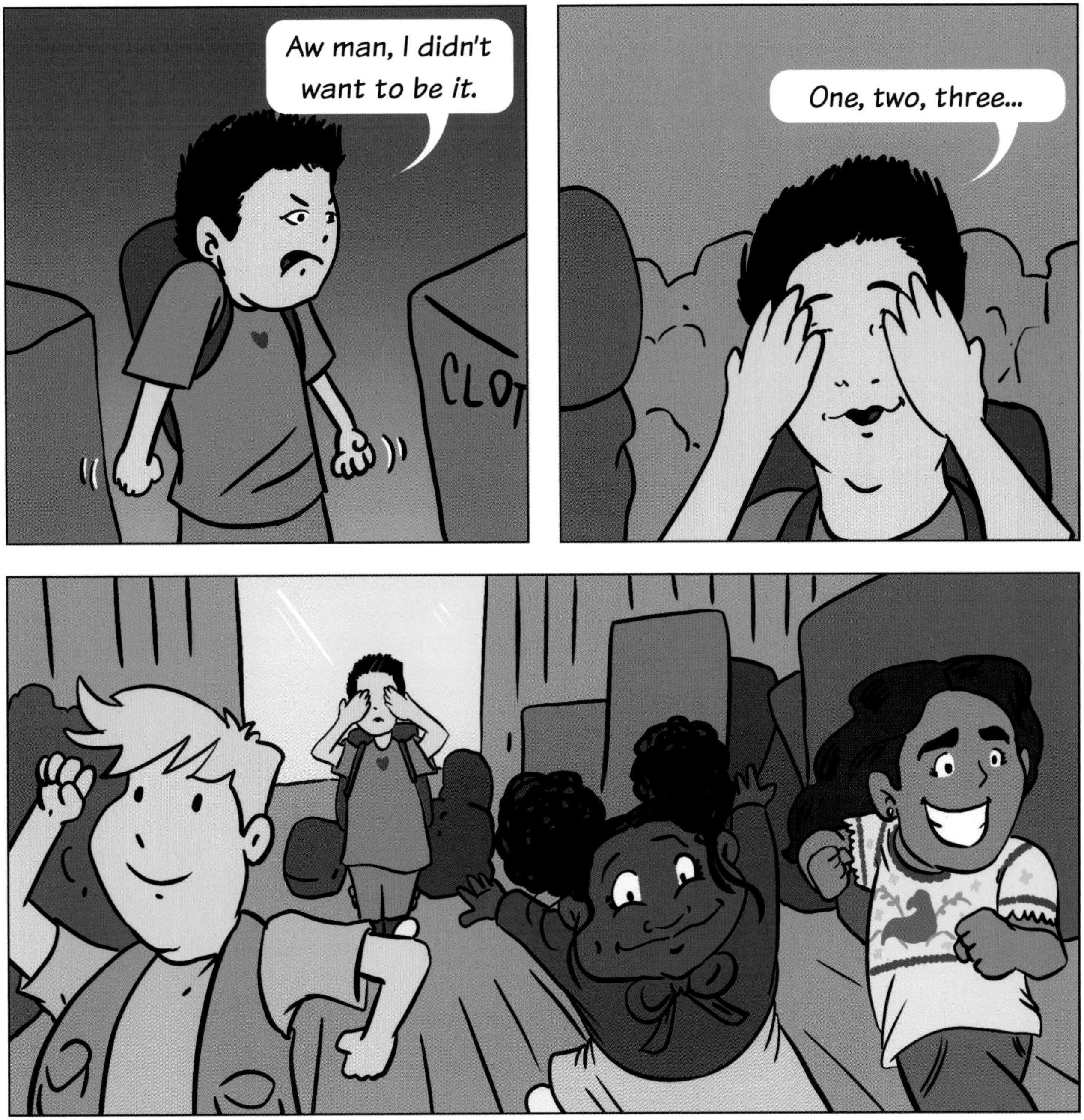
Aw man, I didn't want to be it.
CLOT
One, two, three...

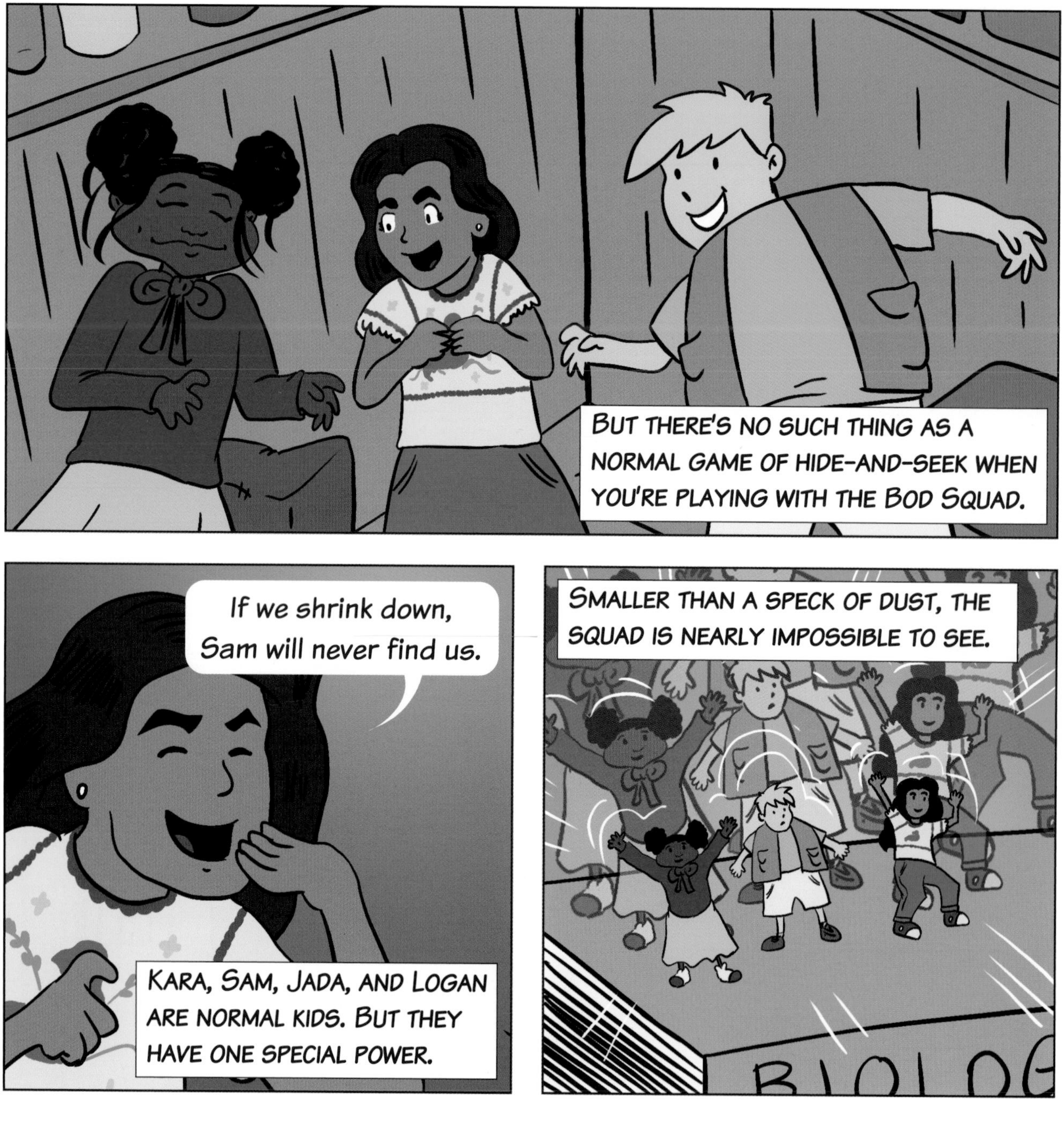
BUT THERE'S NO SUCH THING AS A NORMAL GAME OF HIDE-AND-SEEK WHEN YOU'RE PLAYING WITH THE BOD SQUAD.
If we shrink down, Sam will never find us.
KARA, SAM, JADA, AND LOGAN ARE NORMAL KIDS. BUT THEY HAVE ONE SPECIAL POWER.
SMALLER THAN A SPECK OF DUST, THE SQUAD IS NEARLY IMPOSSIBLE TO SEE.
BIOLOG

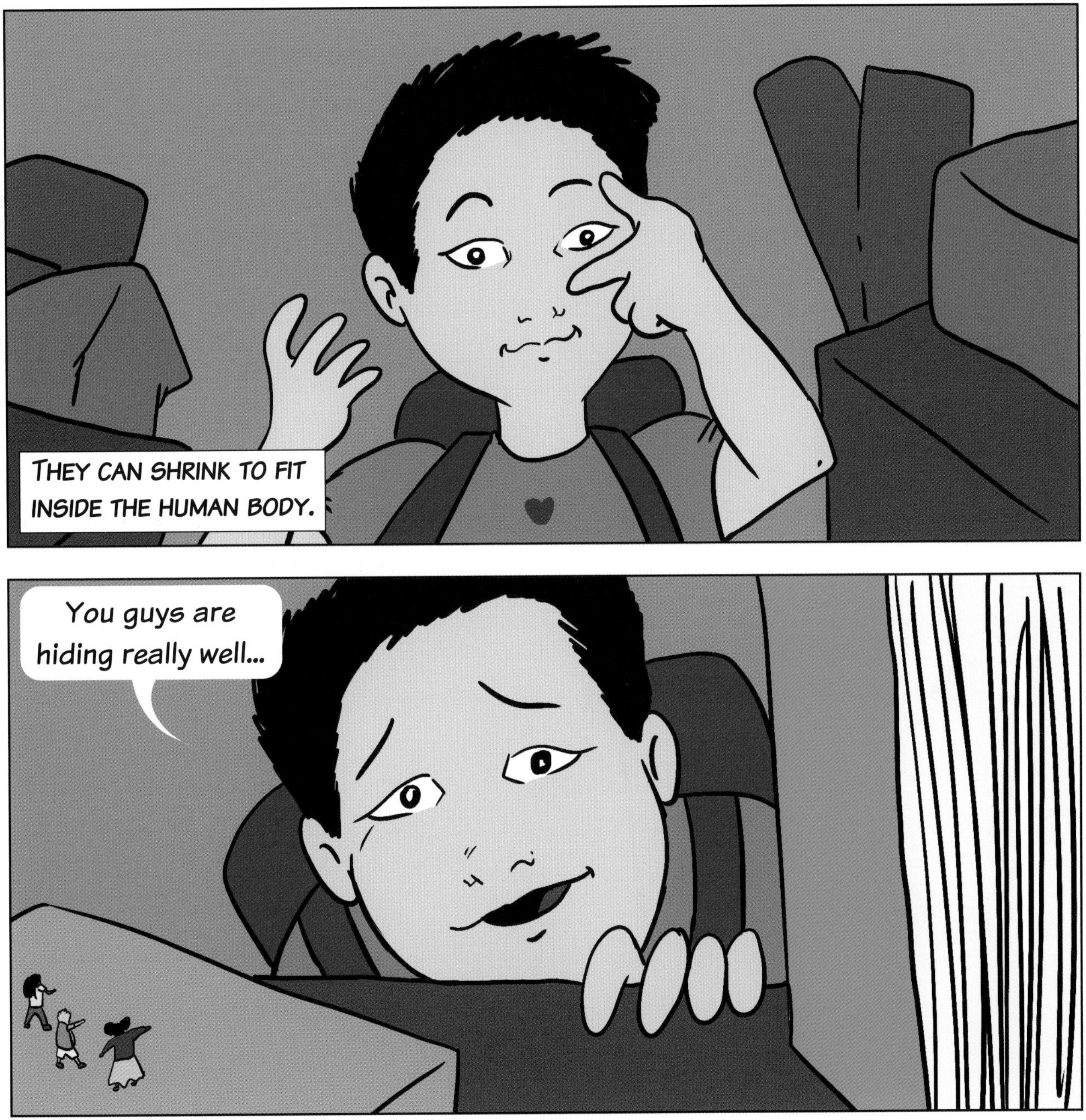
THEY CAN SHRINK TO FIT INSIDE THE HUMAN BODY.
You guys are hiding really well...

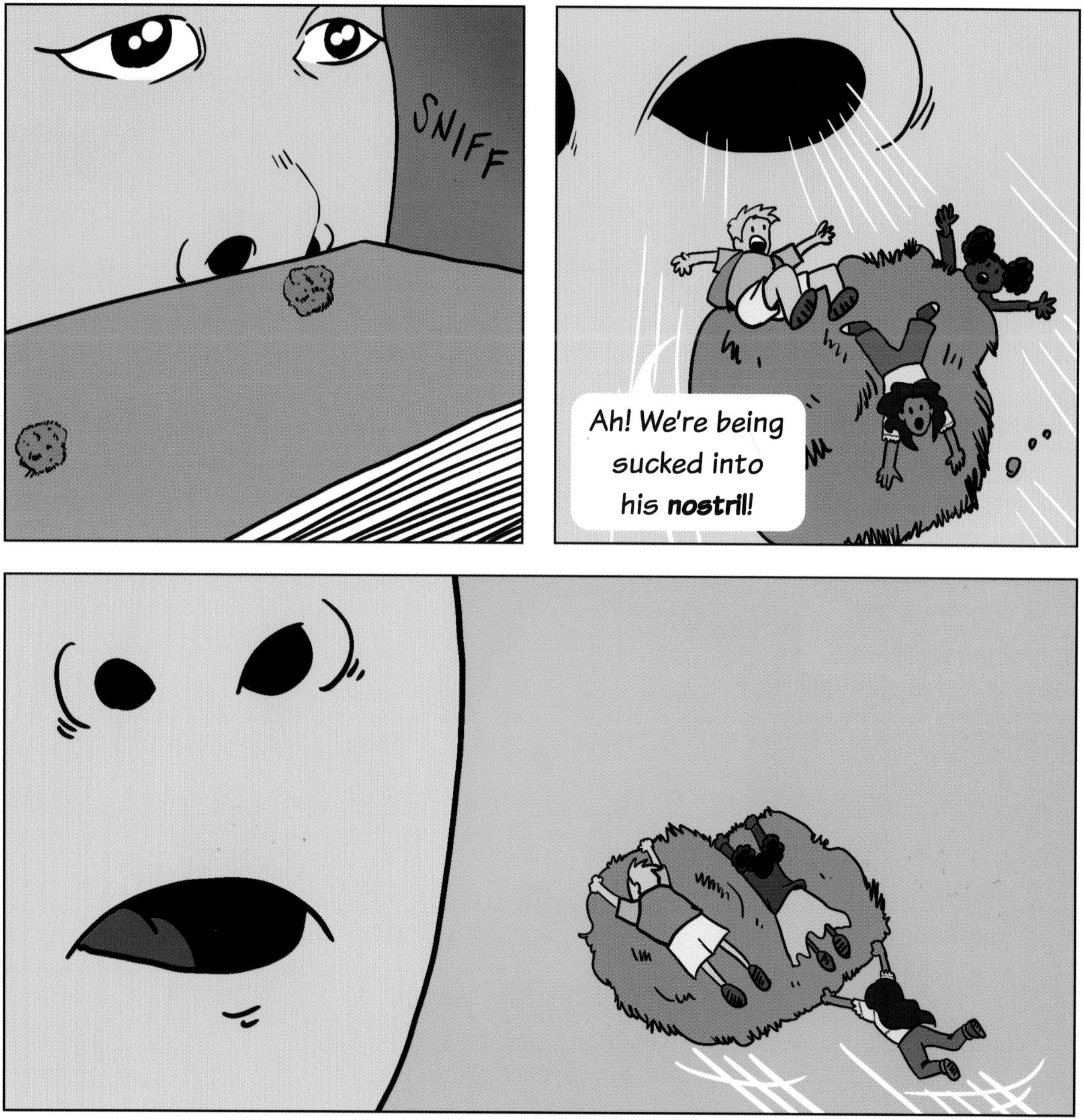
SNIFF
Ah! We're being sucked into his **nostril**!

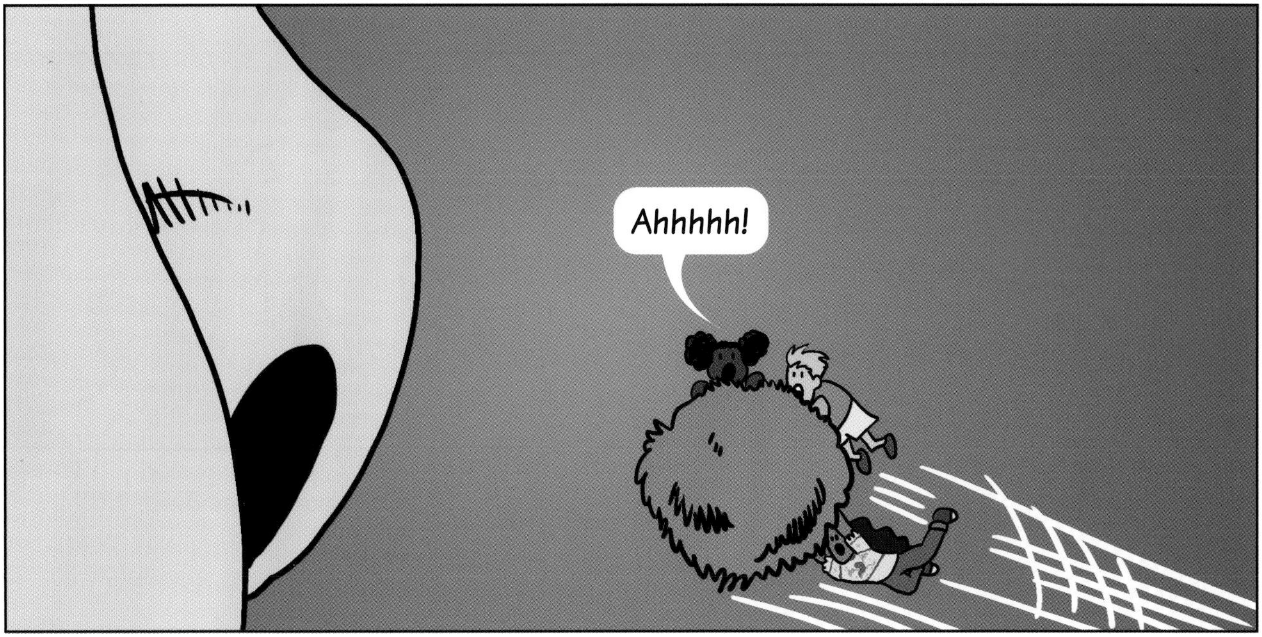
Ahhhhh!

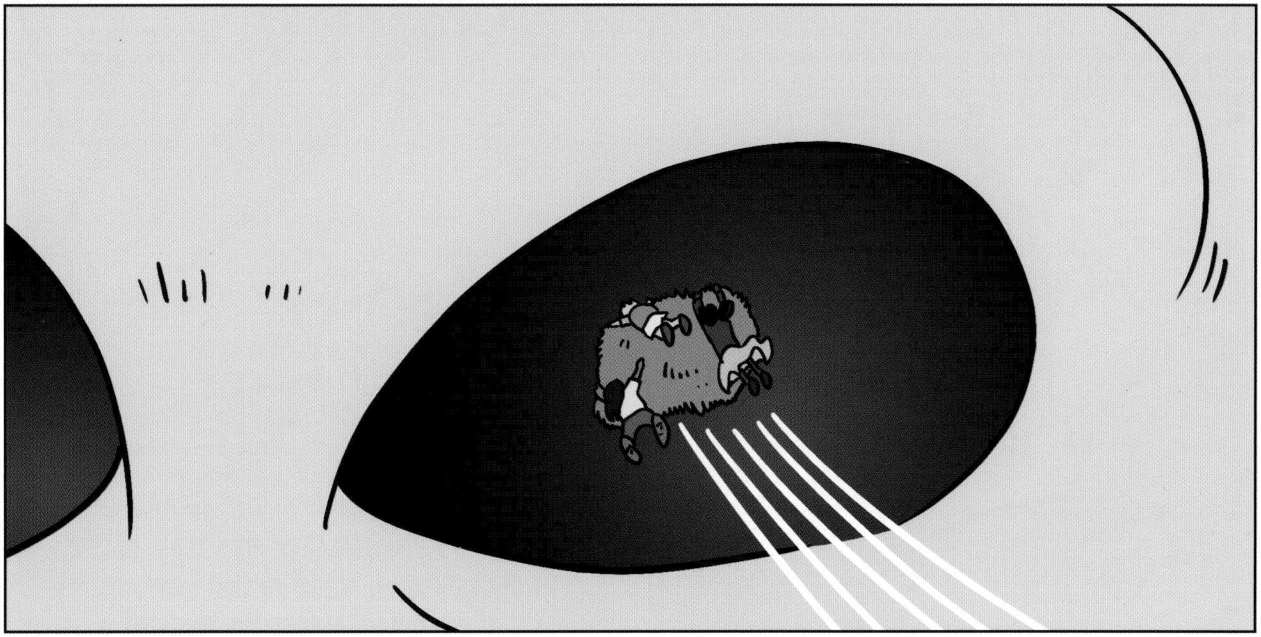

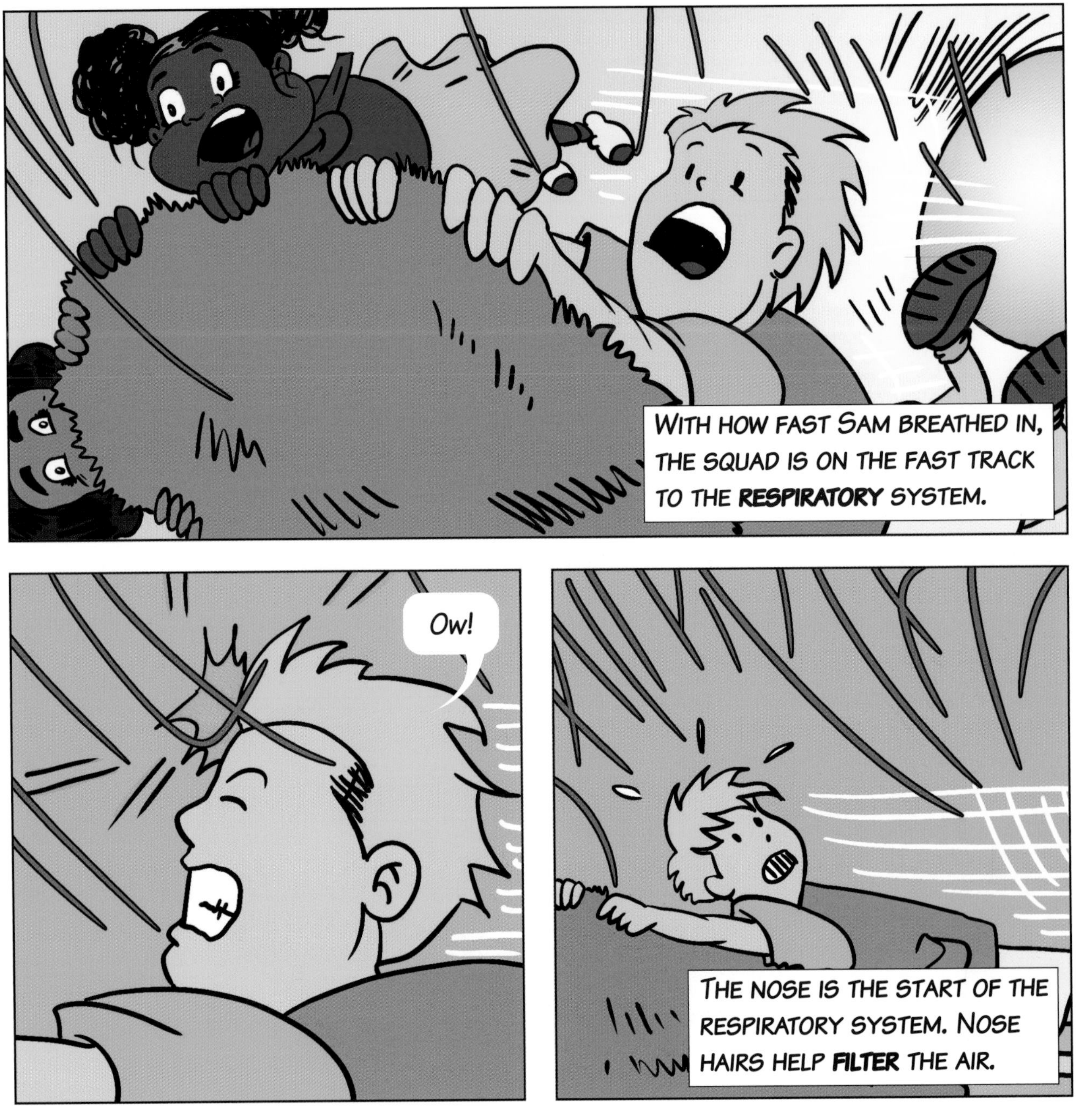
WITH HOW FAST SAM BREATHED IN, THE SQUAD IS ON THE FAST TRACK TO THE **RESPIRATORY** SYSTEM.
Ow!
THE NOSE IS THE START OF THE RESPIRATORY SYSTEM. NOSE HAIRS HELP **FILTER** THE AIR.

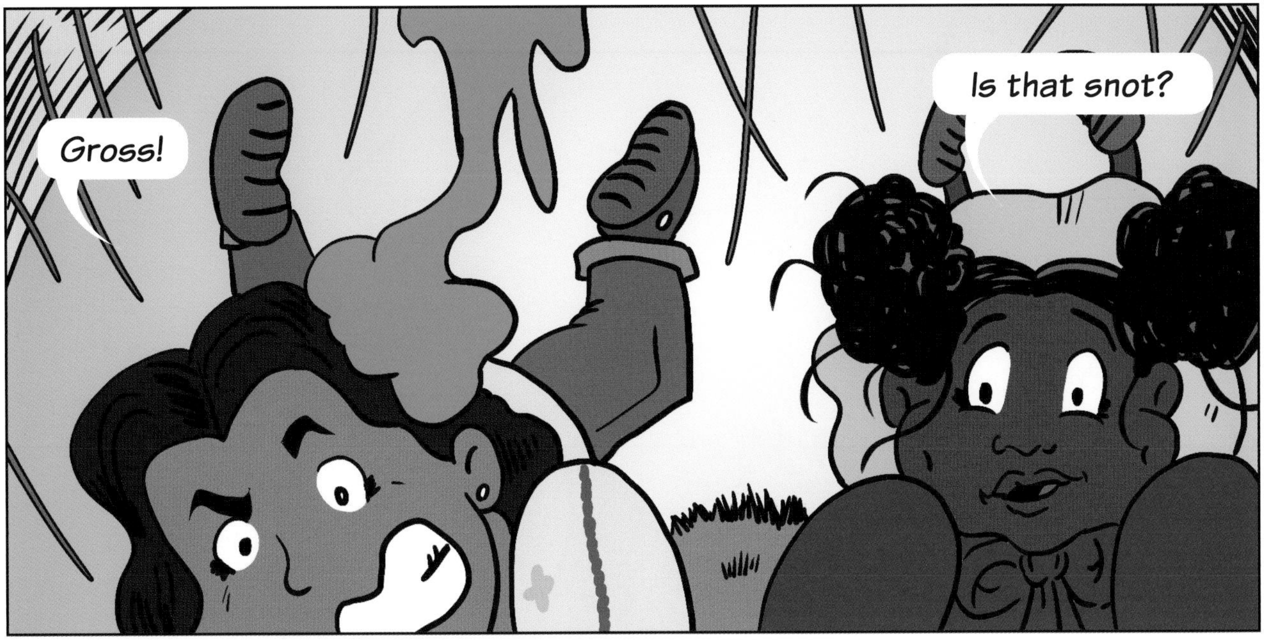
Gross!
Is that snot?

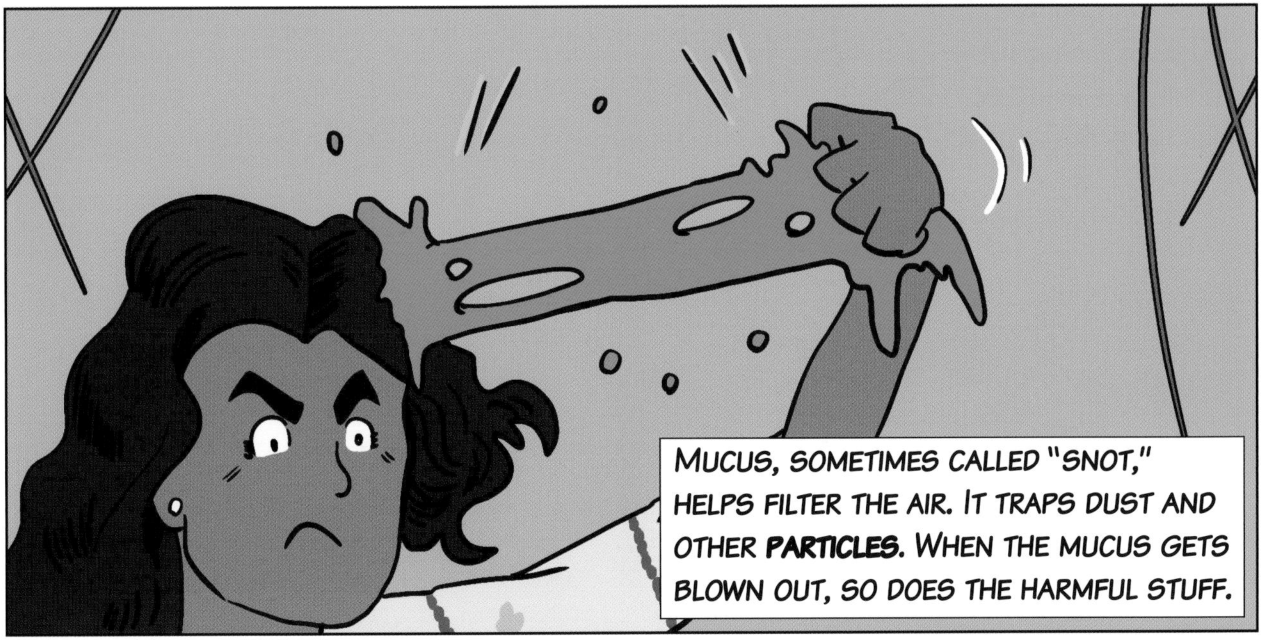
MUCUS, SOMETIMES CALLED "SNOT," HELPS FILTER THE AIR. IT TRAPS DUST AND OTHER **PARTICLES**. WHEN THE MUCUS GETS BLOWN OUT, SO DOES THE HARMFUL STUFF.

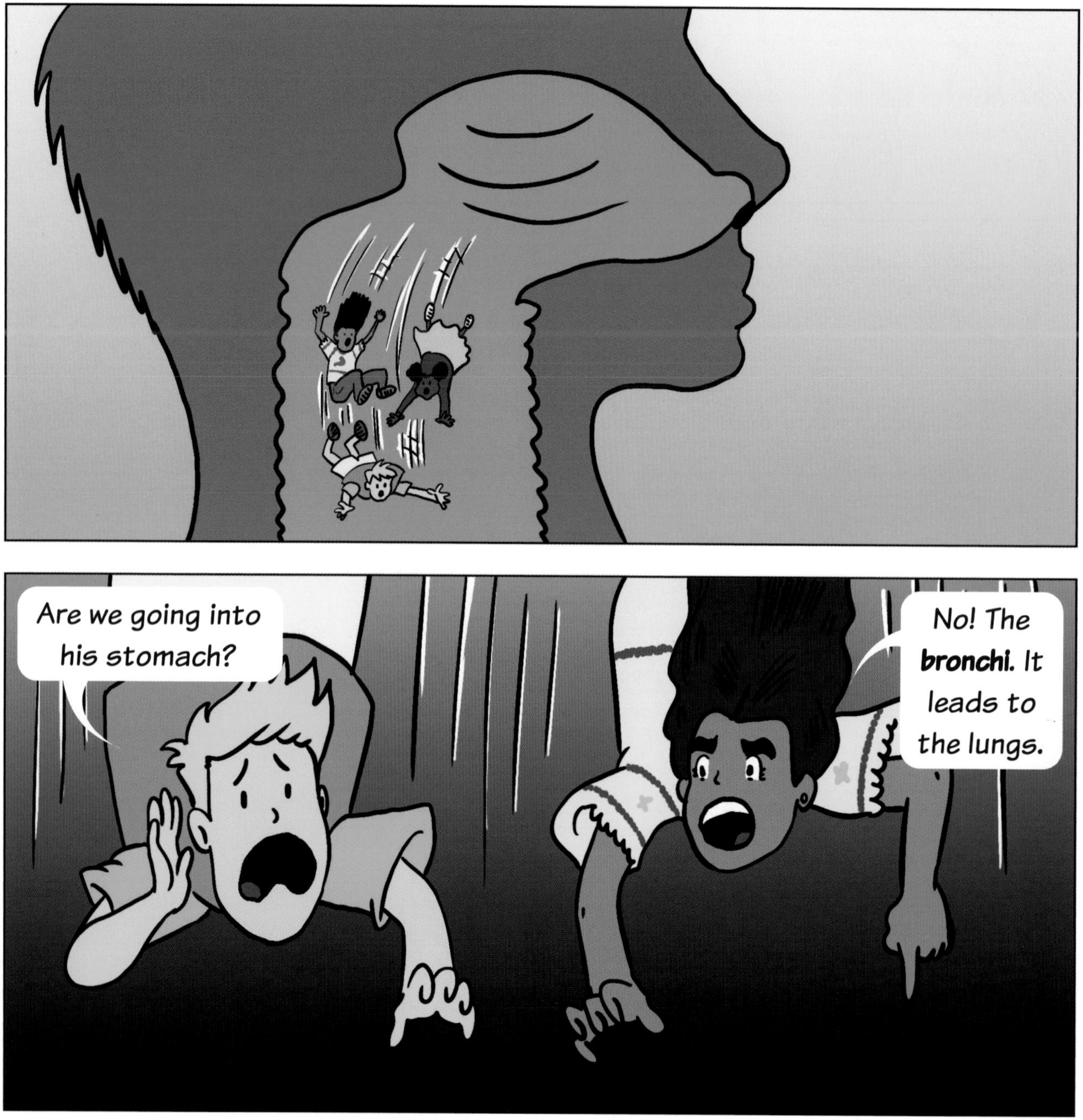
Are we going into his stomach?
No! The **bronchi**. It leads to the lungs.

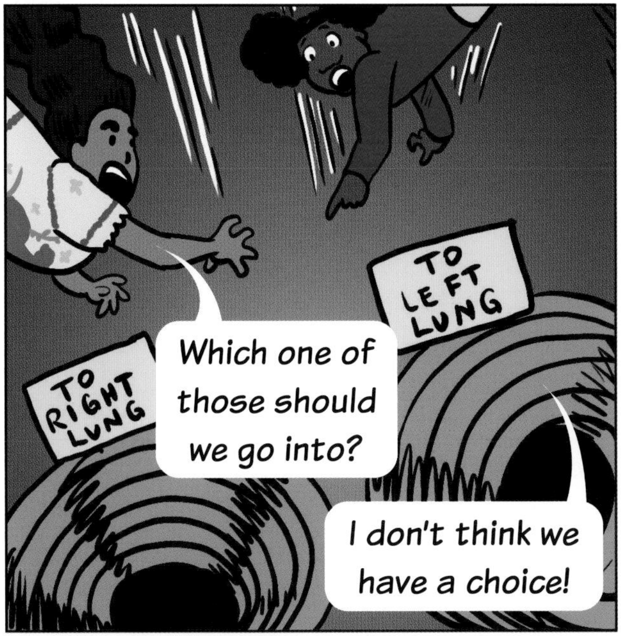
TO LEFT LUNG
TO RIGHT LUNG
Which one of those should we go into?
I don't think we have a choice!

The Lobes of the Lungs

The left and right lungs are not totally the same. The right lung has three lobes. The left lung has two lobes. These are spaces split up by thin walls of tissue. The left lung is also a bit smaller. This leaves room for the heart.

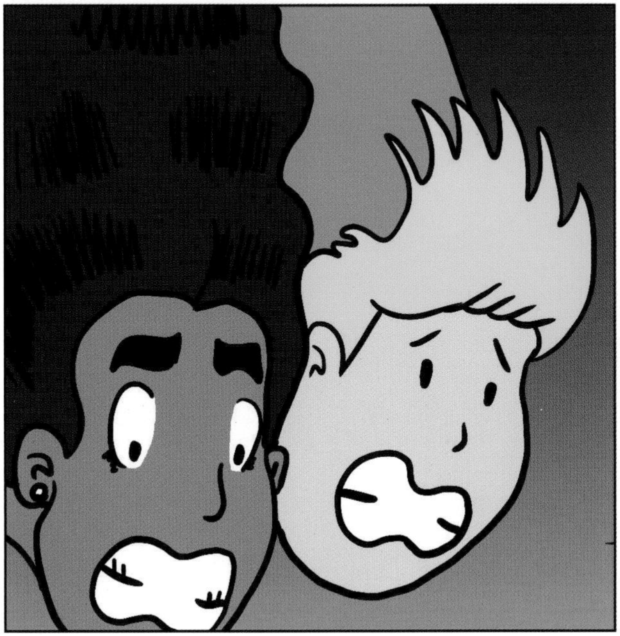

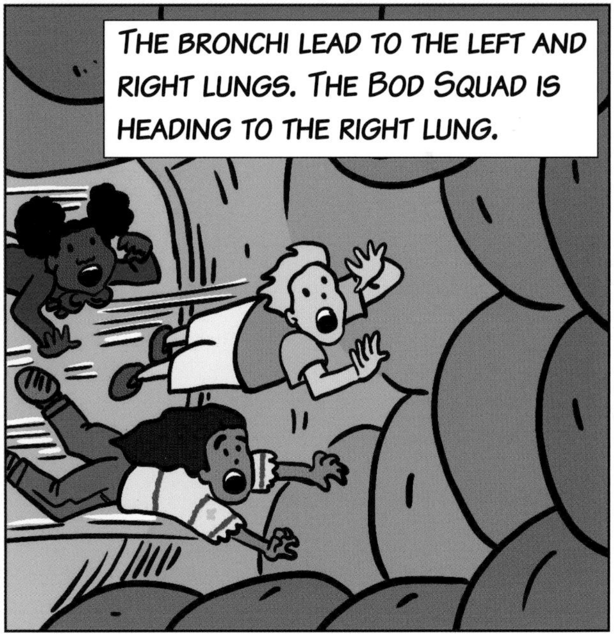
THE BRONCHI LEAD TO THE LEFT AND RIGHT LUNGS. THE BOD SQUAD IS HEADING TO THE RIGHT LUNG.

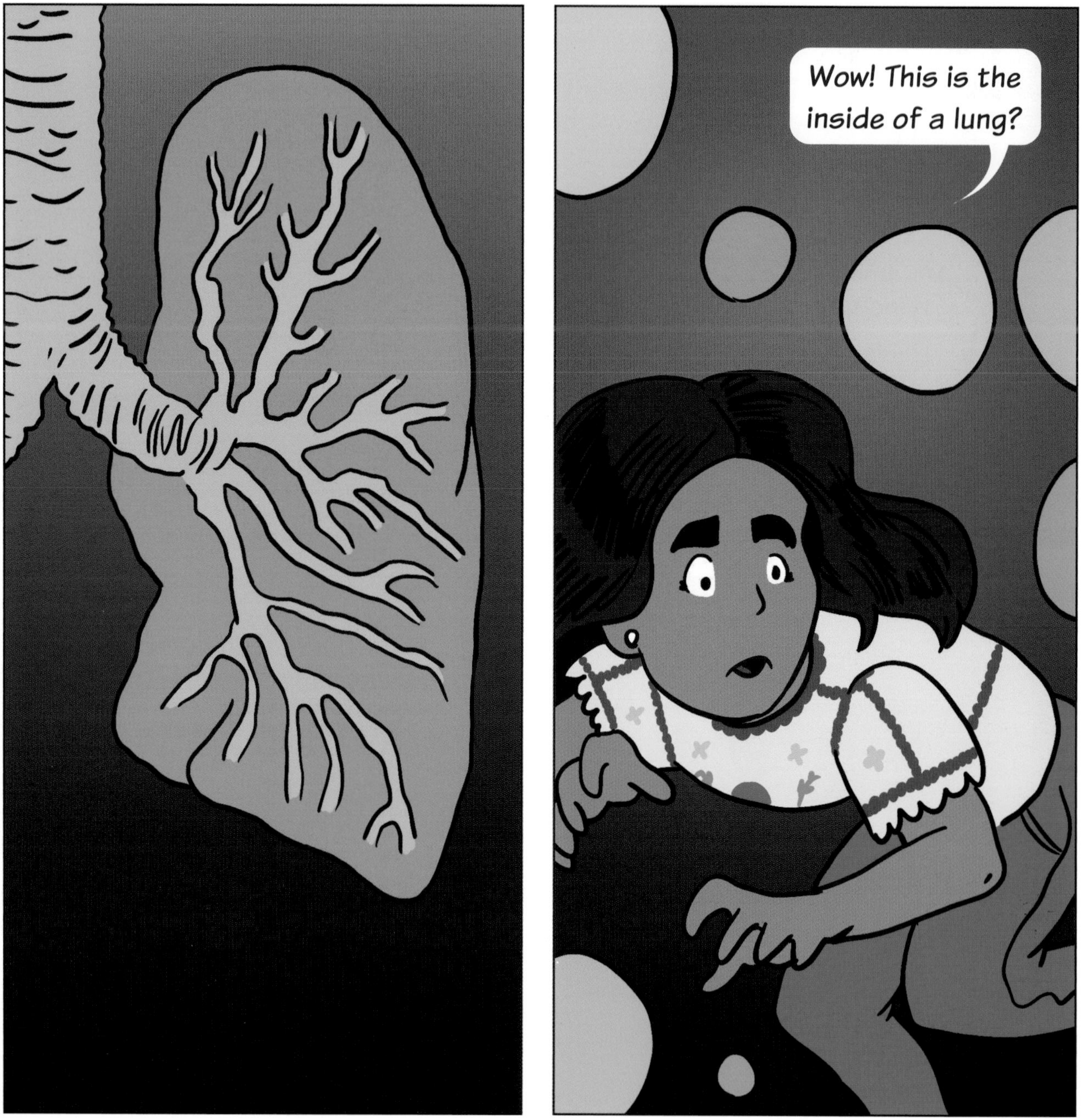
Wow! This is the inside of a lung?

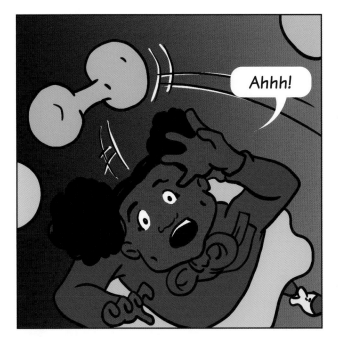

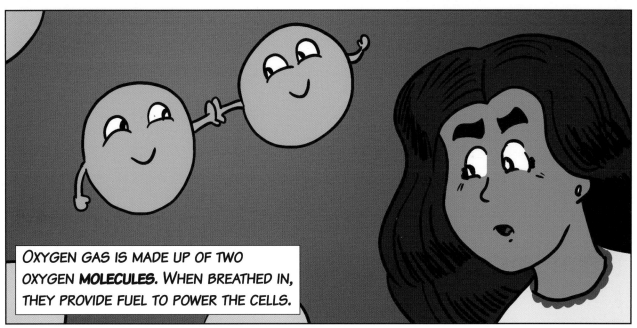

Oxygen gas is made up of two oxygen **molecules**. When breathed in, they provide fuel to power the cells.

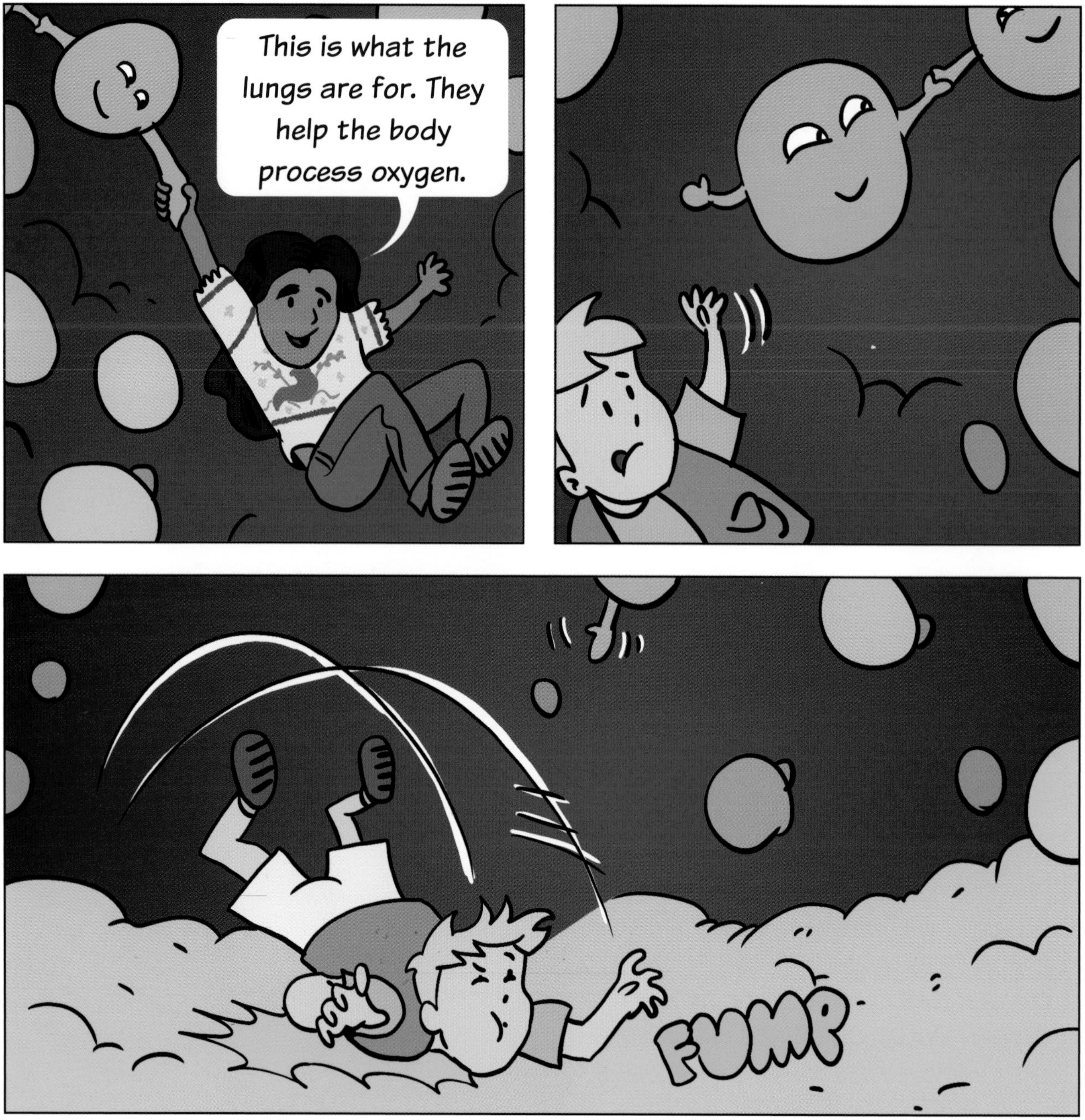
This is what the lungs are for. They help the body process oxygen.
FUMP

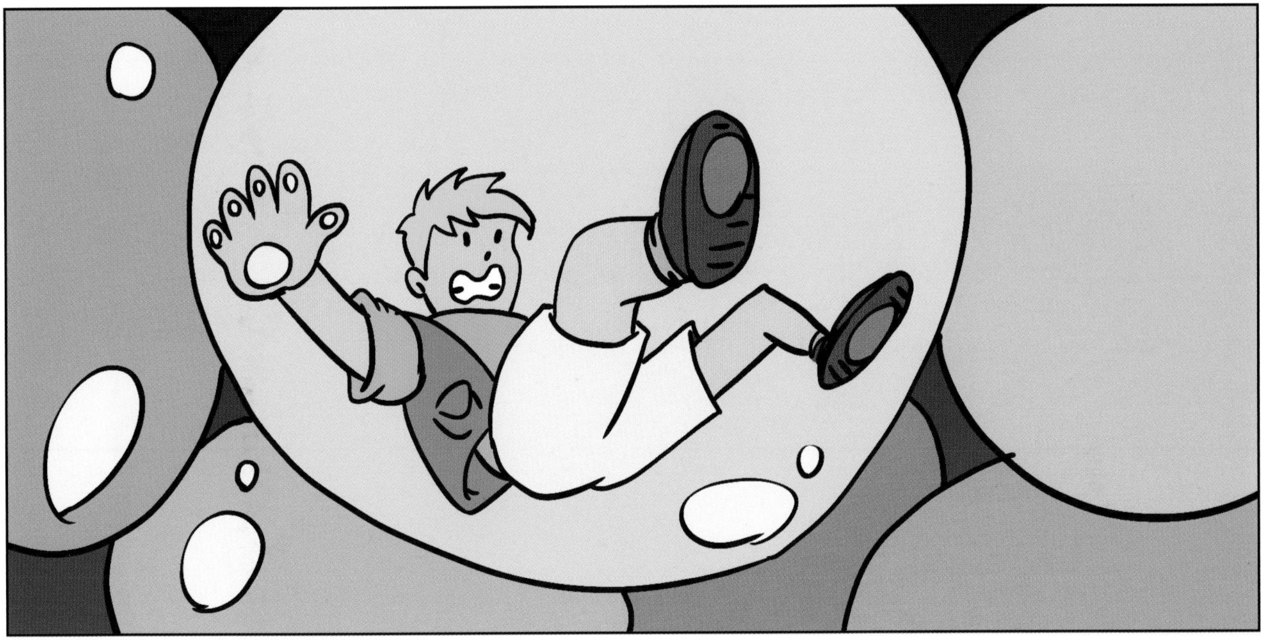

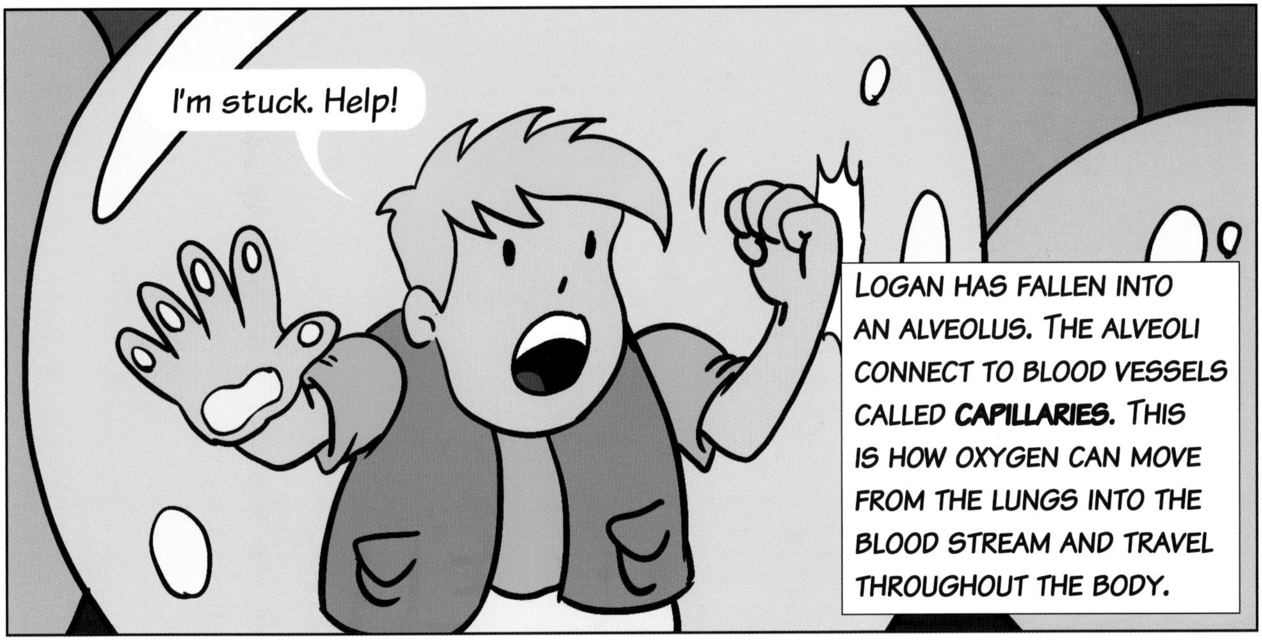
I'm stuck. Help!
LOGAN HAS FALLEN INTO AN ALVEOLUS. THE ALVEOLI CONNECT TO BLOOD VESSELS CALLED **CAPILLARIES.** THIS IS HOW OXYGEN CAN MOVE FROM THE LUNGS INTO THE BLOOD STREAM AND TRAVEL THROUGHOUT THE BODY.

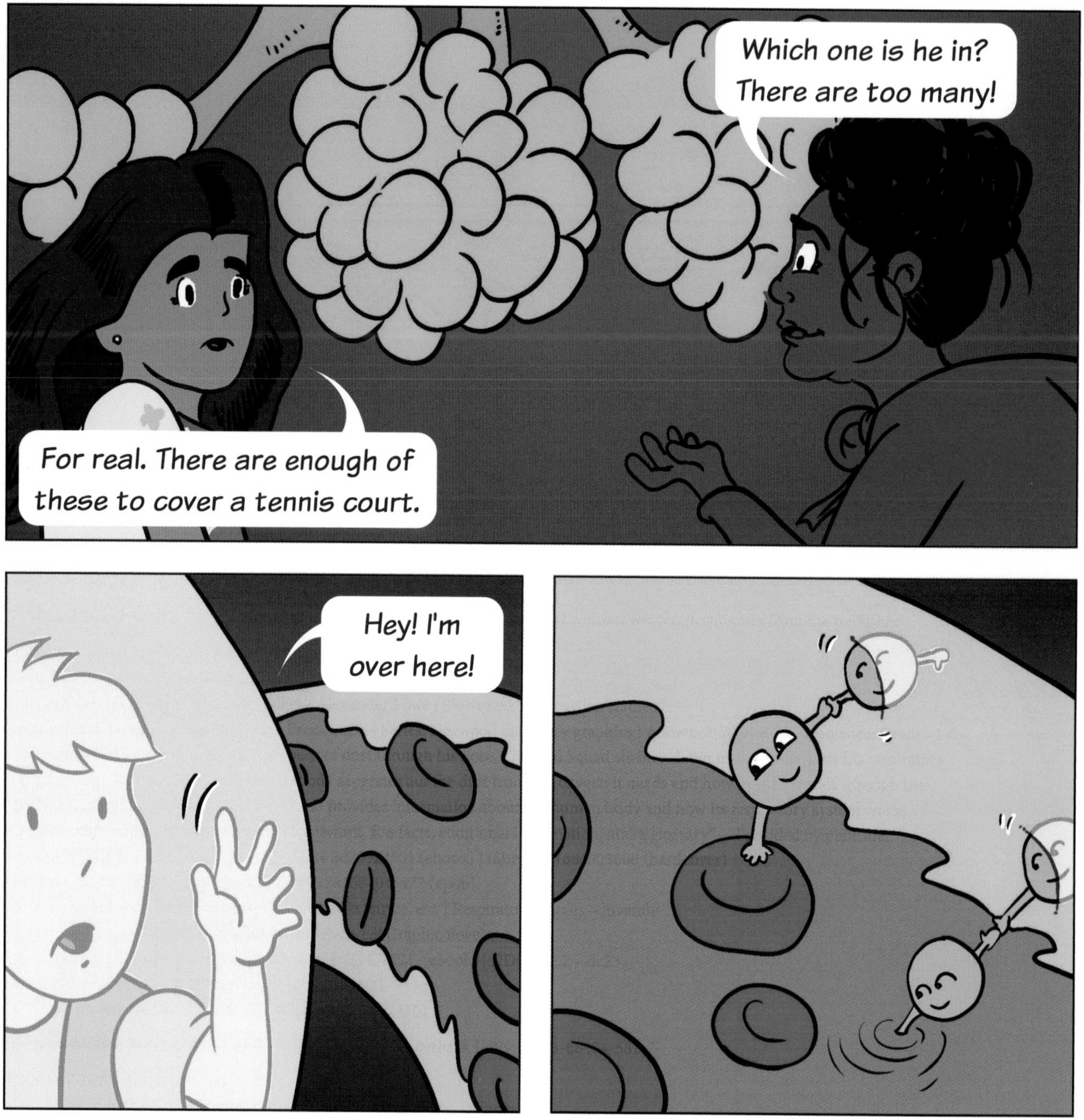
Which one is he in? There are too many!
For real. There are enough of these to cover a tennis court.
Hey! I'm over here!

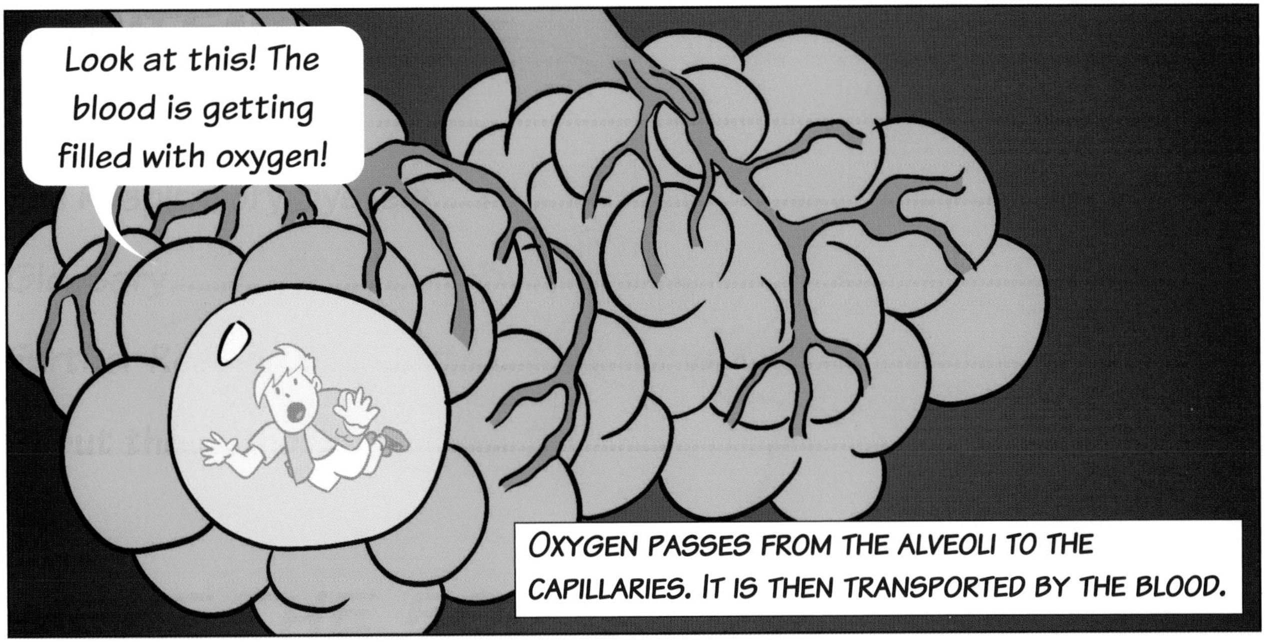
Look at this! The blood is getting filled with oxygen!
OXYGEN PASSES FROM THE ALVEOLI TO THE CAPILLARIES. IT IS THEN TRANSPORTED BY THE BLOOD.

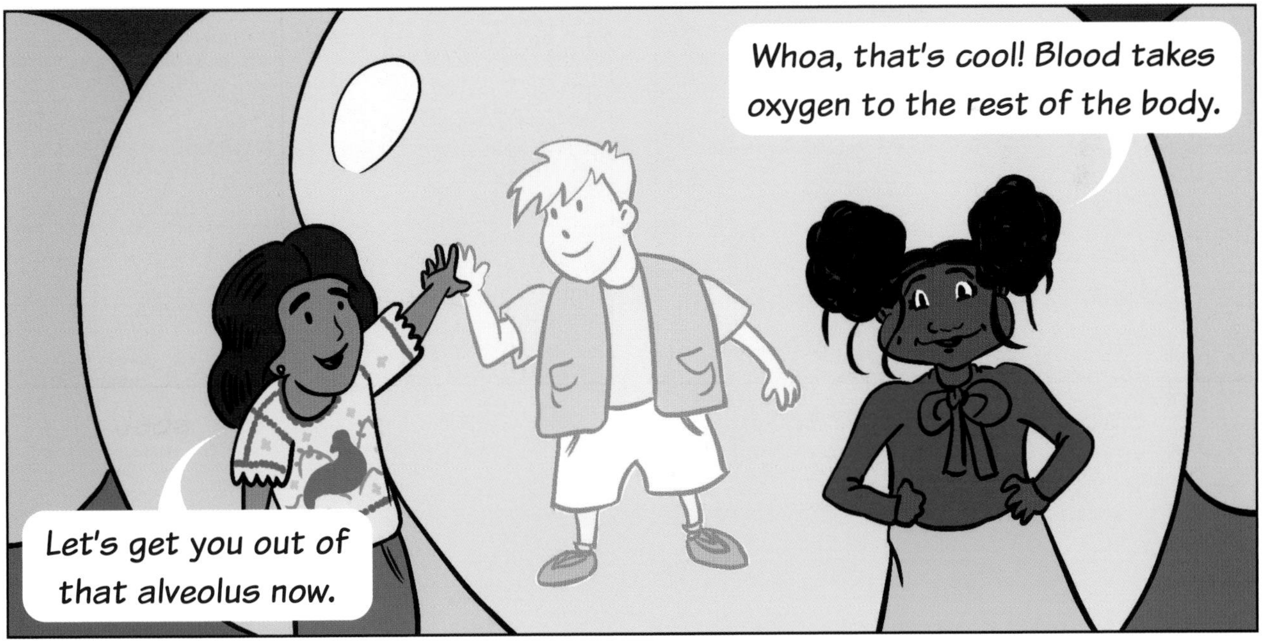
Whoa, that's cool! Blood takes oxygen to the rest of the body.
Let's get you out of that alveolus now.

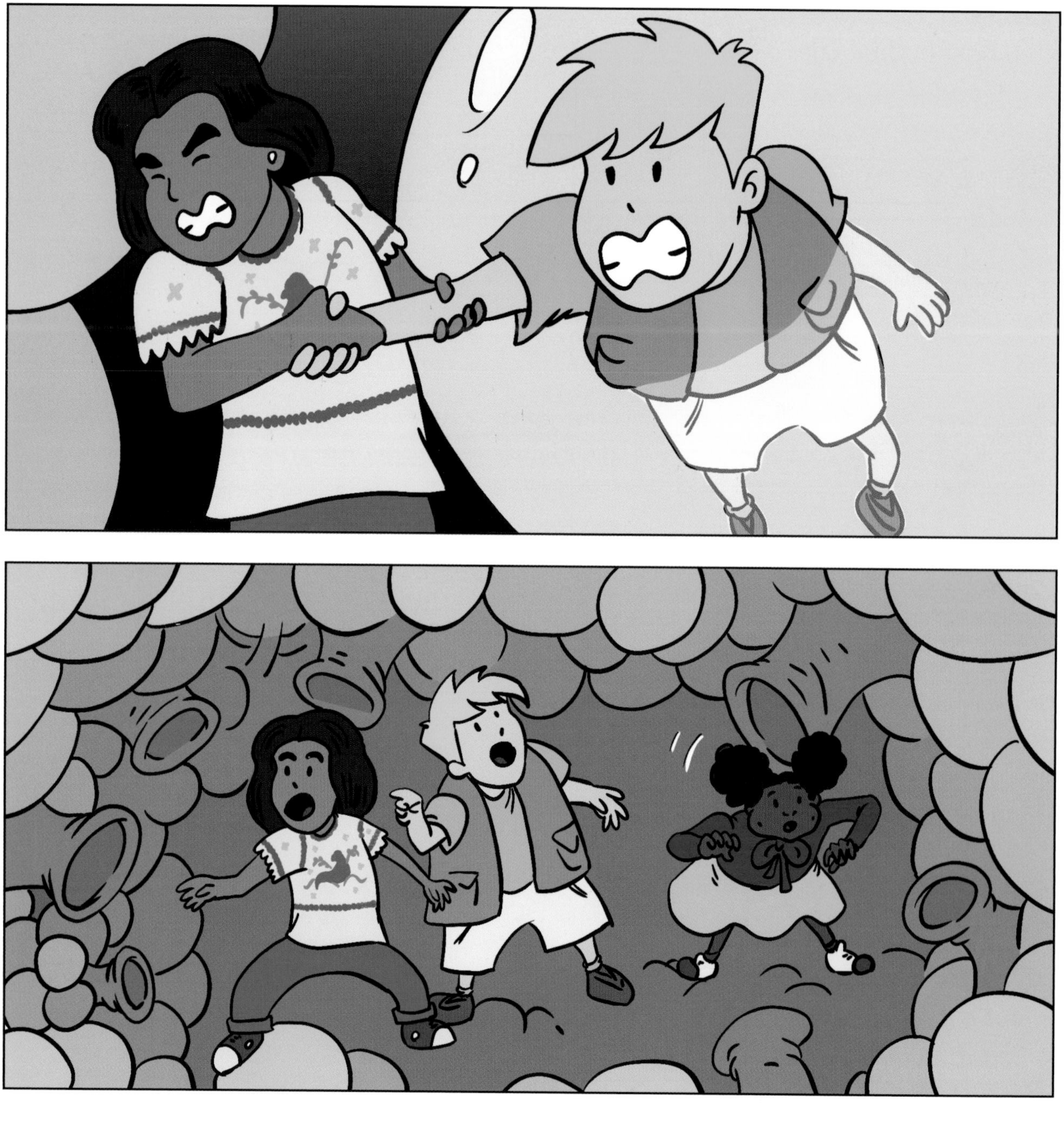

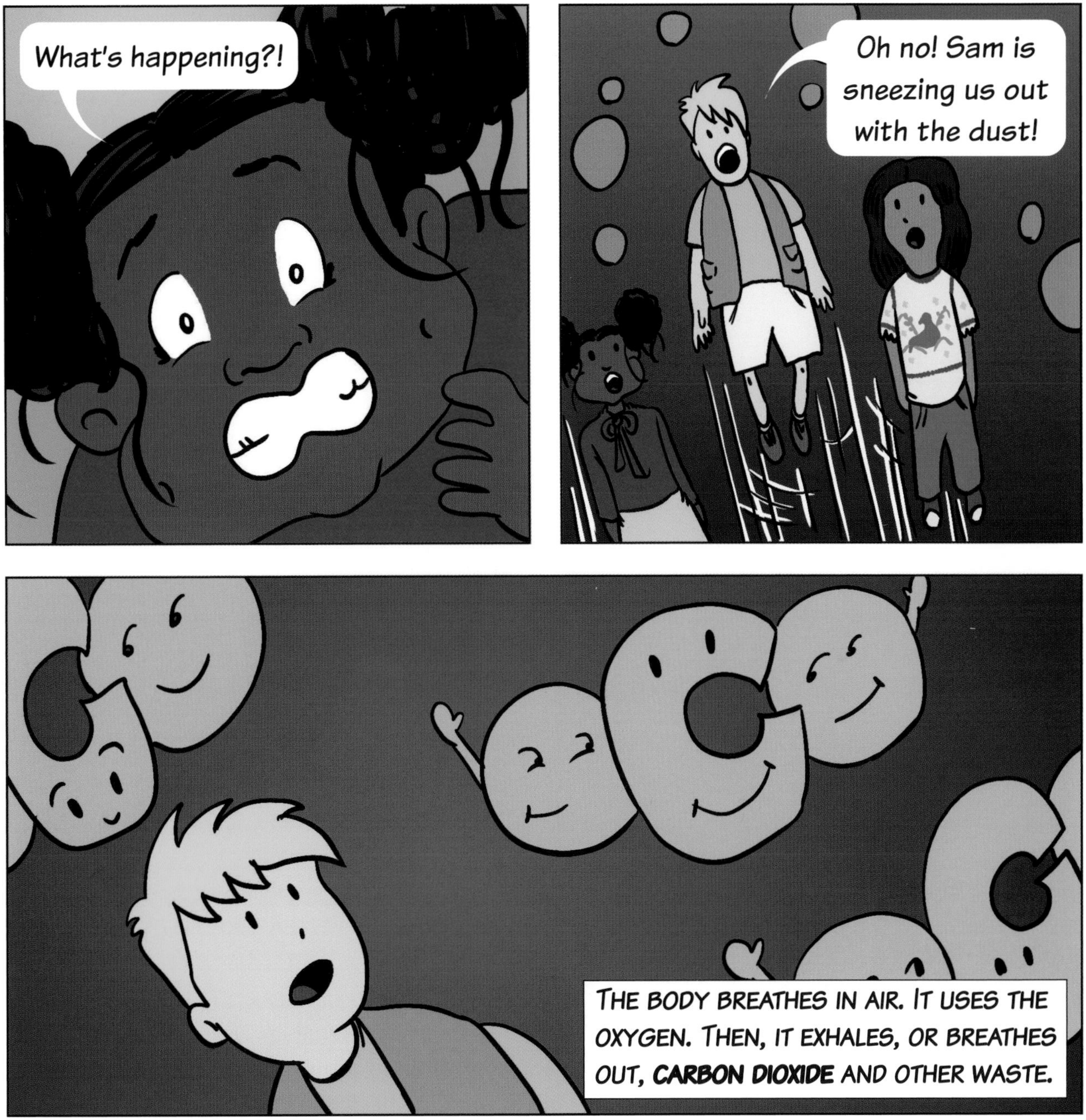
What's happening?!
Oh no! Sam is sneezing us out with the dust!
THE BODY BREATHES IN AIR. IT USES THE OXYGEN. THEN, IT EXHALES, OR BREATHES OUT, **CARBON DIOXIDE** AND OTHER WASTE.

This is way too fast!

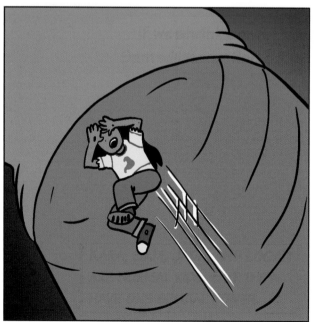

THE DIAPHRAGM IS A MUSCLE BELOW THE LUNGS AND HEART. IT HELPS BREATHE AIR IN. WHEN IT RELAXES, AIR IS PUSHED OUT.

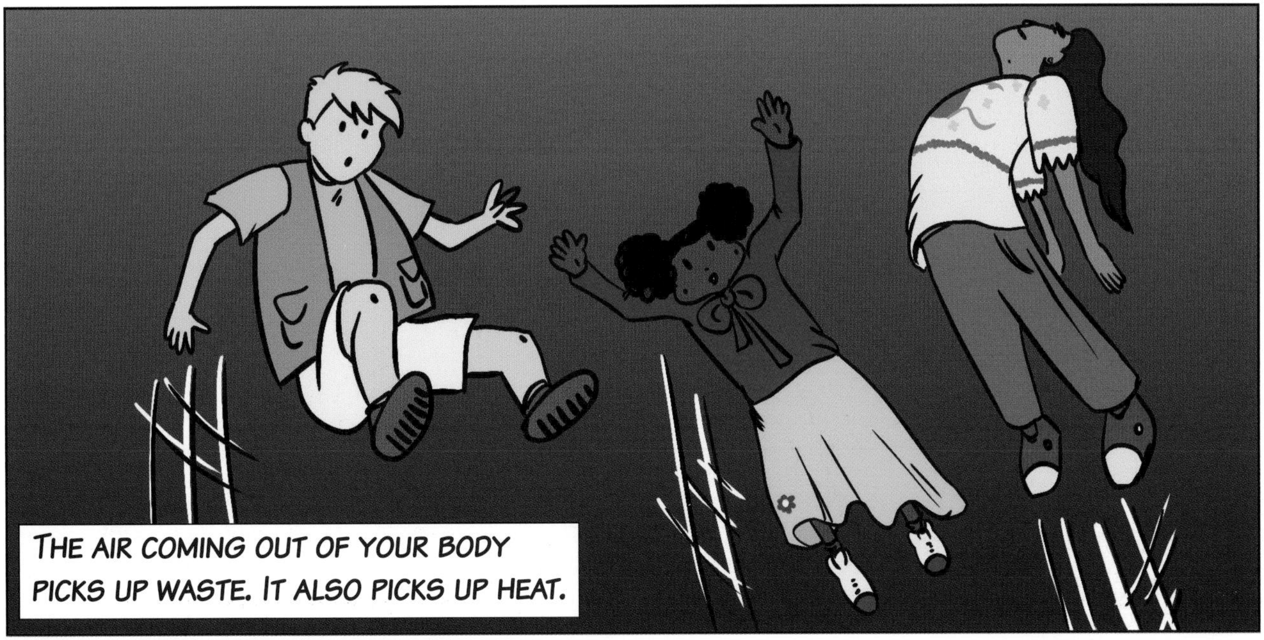
THE AIR COMING OUT OF YOUR BODY PICKS UP WASTE. IT ALSO PICKS UP HEAT.

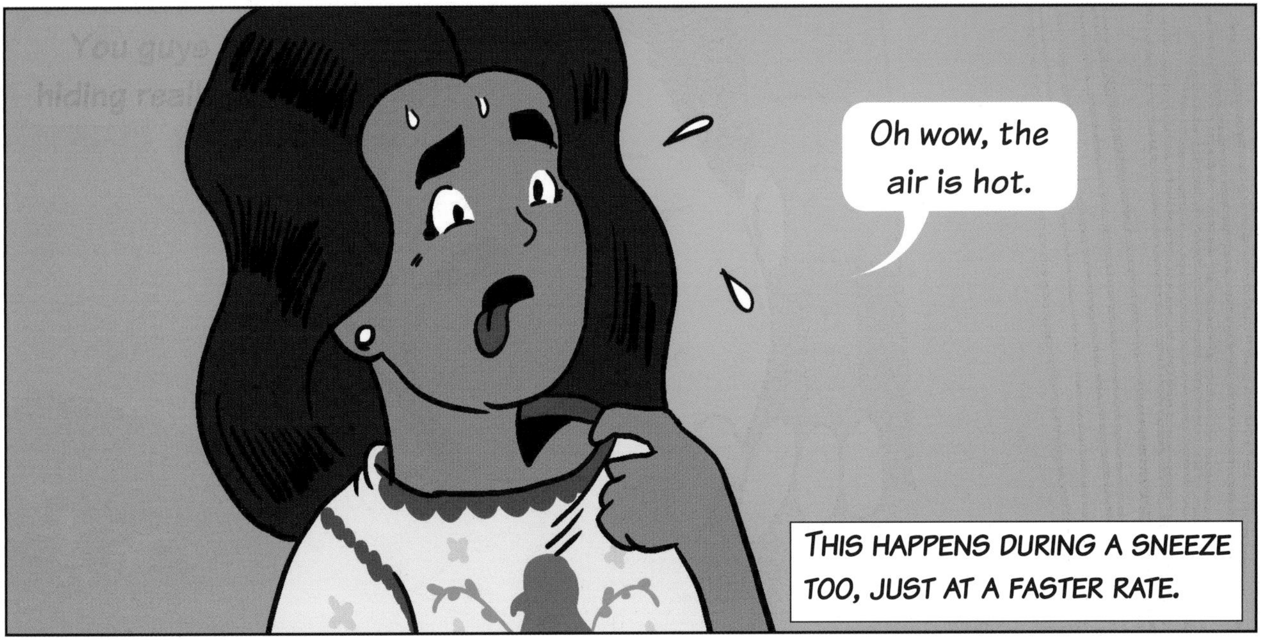
Oh wow, the air is hot.
THIS HAPPENS DURING A SNEEZE TOO, JUST AT A FASTER RATE.

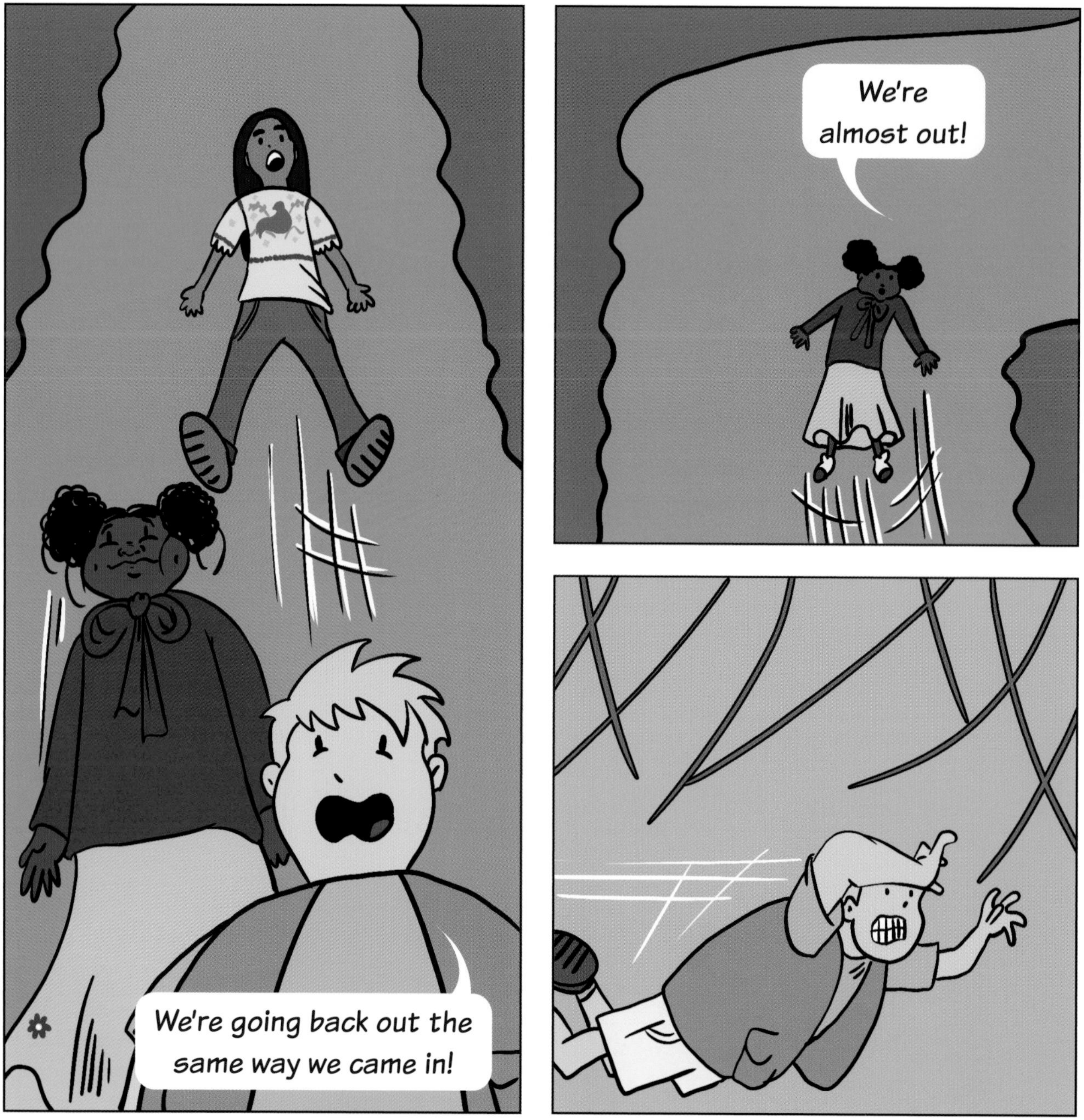
We're going back out the same way we came in!
We're almost out!

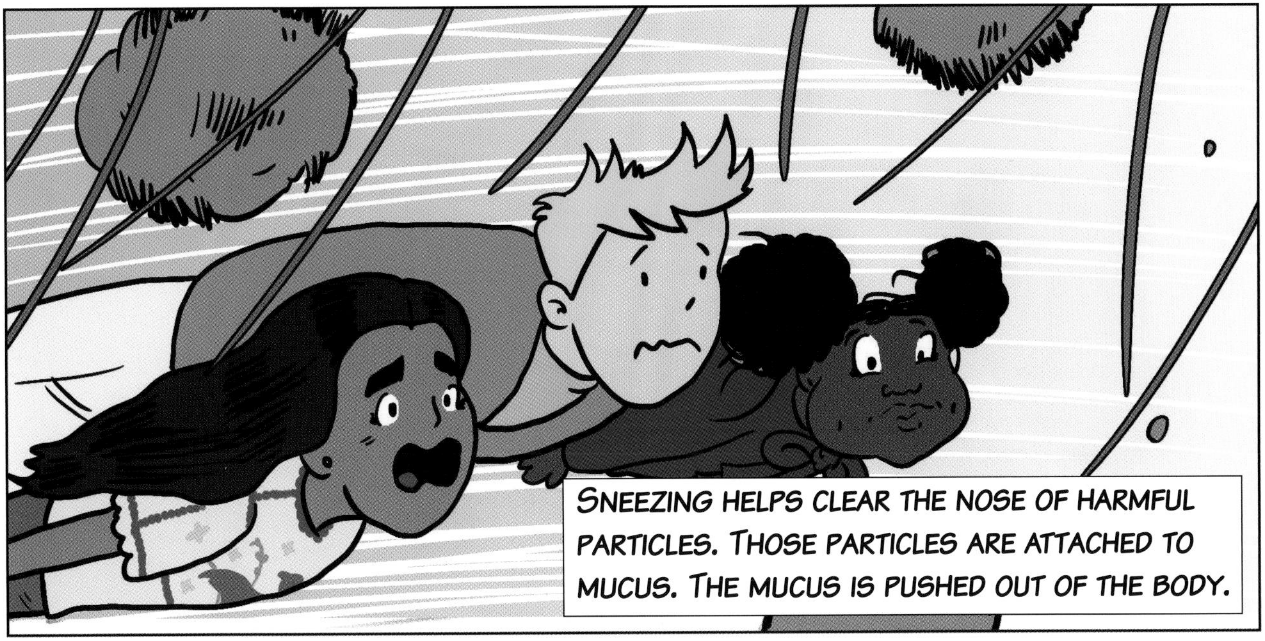
SNEEZING HELPS CLEAR THE NOSE OF HARMFUL PARTICLES. THOSE PARTICLES ARE ATTACHED TO MUCUS. THE MUCUS IS PUSHED OUT OF THE BODY.

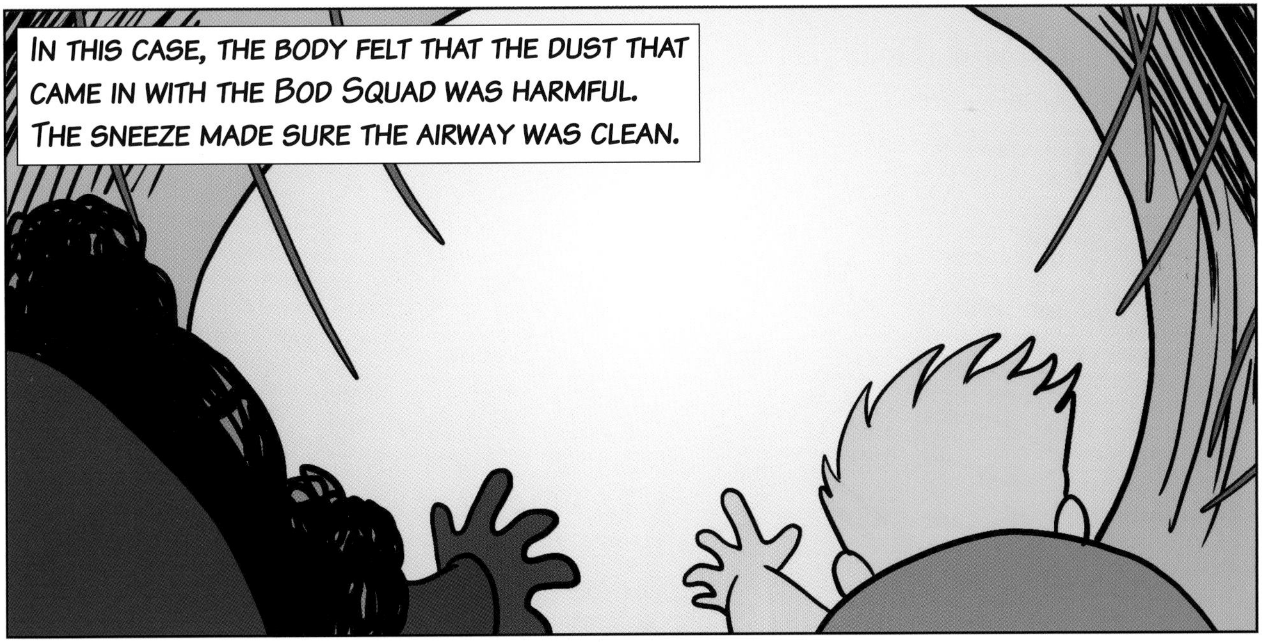
IN THIS CASE, THE BODY FELT THAT THE DUST THAT CAME IN WITH THE BOD SQUAD WAS HARMFUL. THE SNEEZE MADE SURE THE AIRWAY WAS CLEAN.

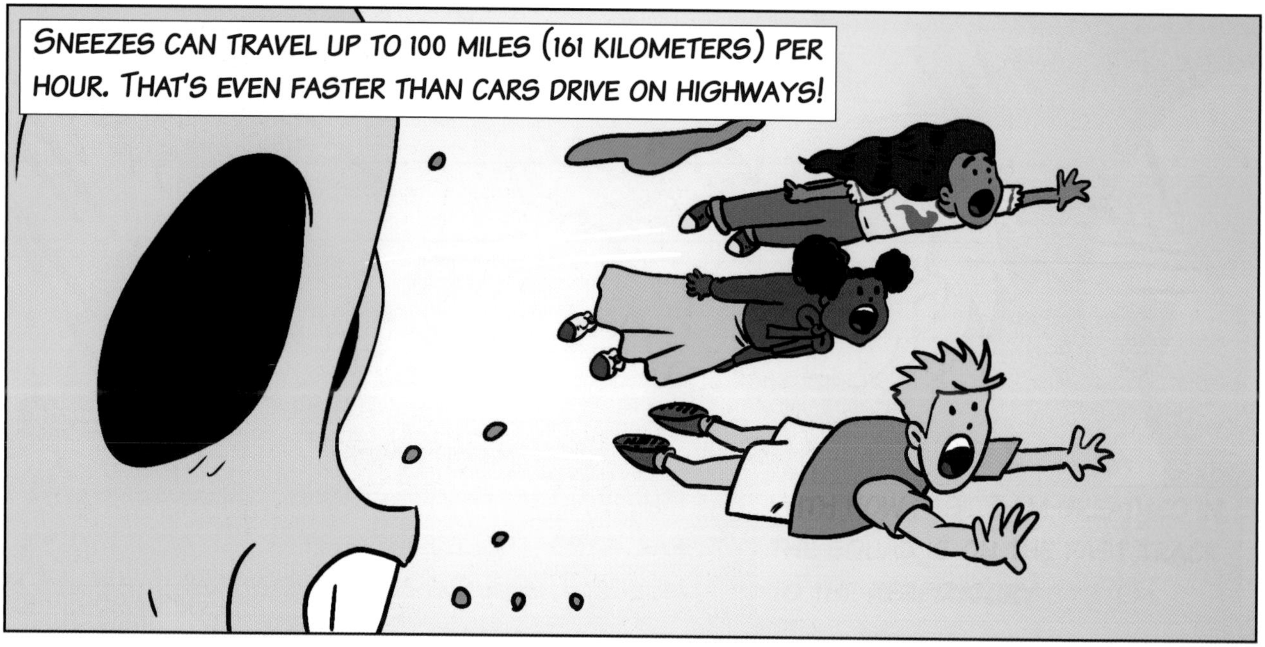
SNEEZES CAN TRAVEL UP TO 100 MILES (161 KILOMETERS) PER HOUR. THAT'S EVEN FASTER THAN CARS DRIVE ON HIGHWAYS!

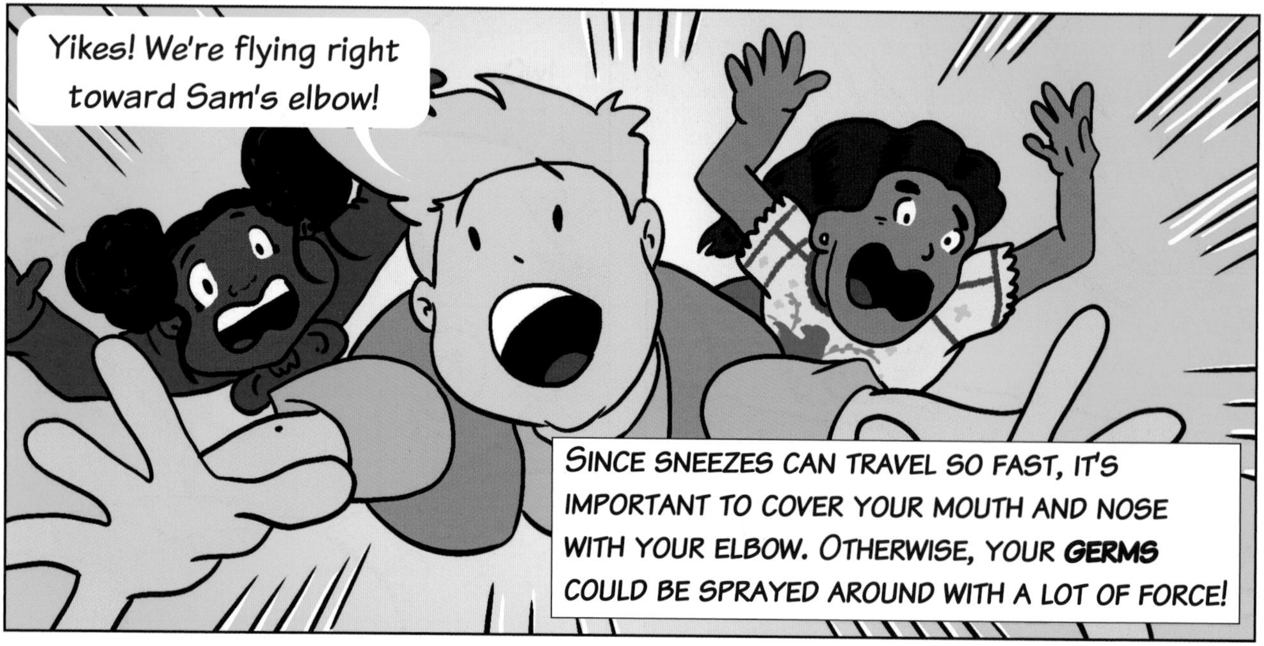
Yikes! We're flying right toward Sam's elbow!
SINCE SNEEZES CAN TRAVEL SO FAST, IT'S IMPORTANT TO COVER YOUR MOUTH AND NOSE WITH YOUR ELBOW. OTHERWISE, YOUR **GERMS** COULD BE SPRAYED AROUND WITH A LOT OF FORCE!

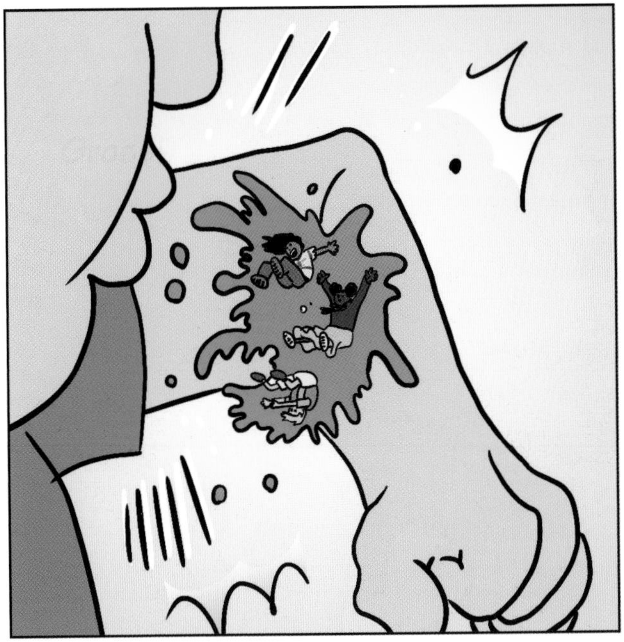

What Is Mucus?

When some people hear about mucus, they think it's only the slimy stuff in their nose. But mucus is all throughout the body. It protects different **organs**. It is almost like a lotion for the inside. A normal body produces 1 to 2 quarts (4 to 8 cups) of mucus a day.

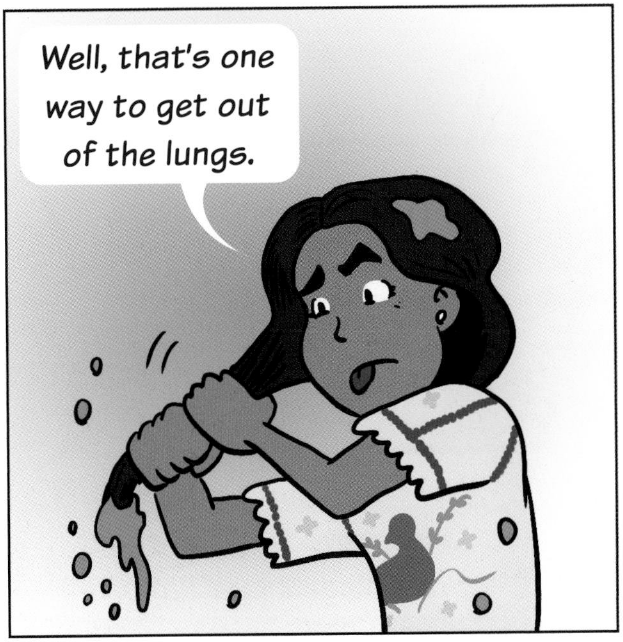
Well, that's one way to get out of the lungs.

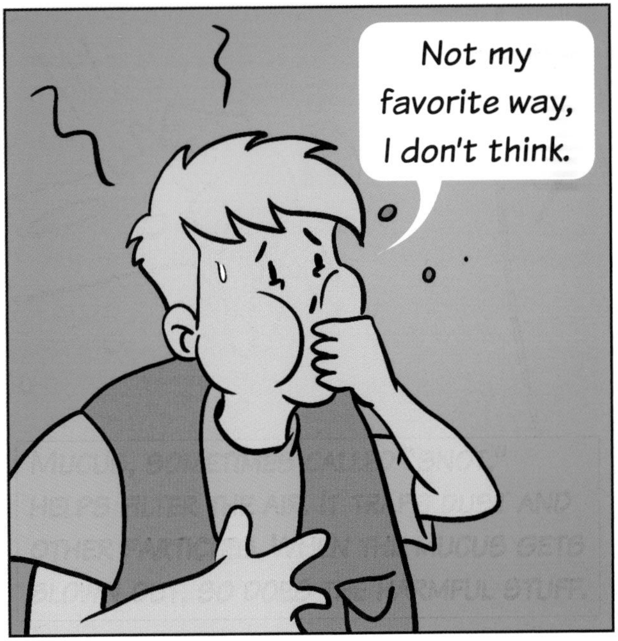
Not my favorite way, I don't think.

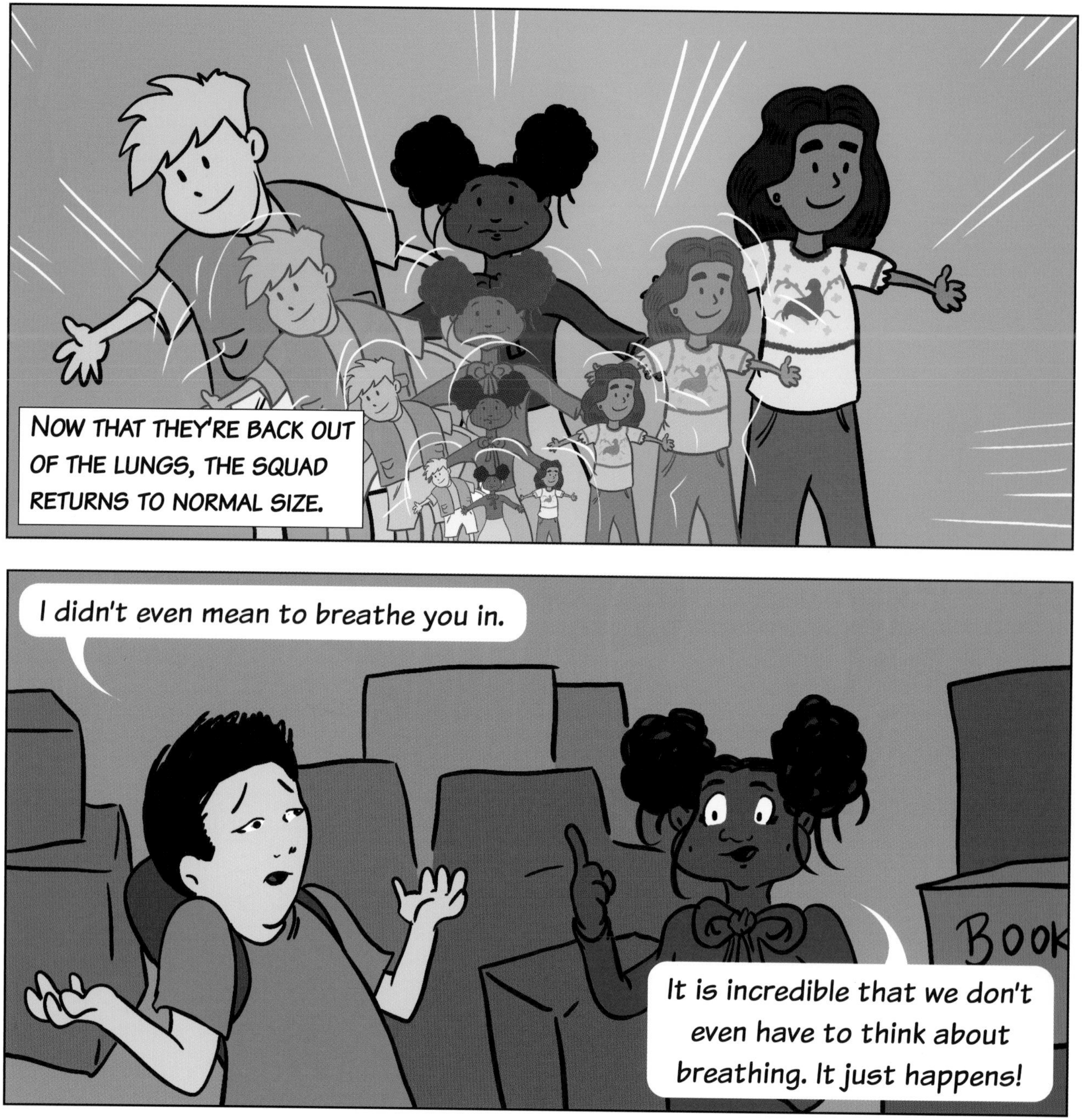
NOW THAT THEY'RE BACK OUT OF THE LUNGS, THE SQUAD RETURNS TO NORMAL SIZE.
I didn't even mean to breathe you in.
BOOK
It is incredible that we don't even have to think about breathing. It just happens!

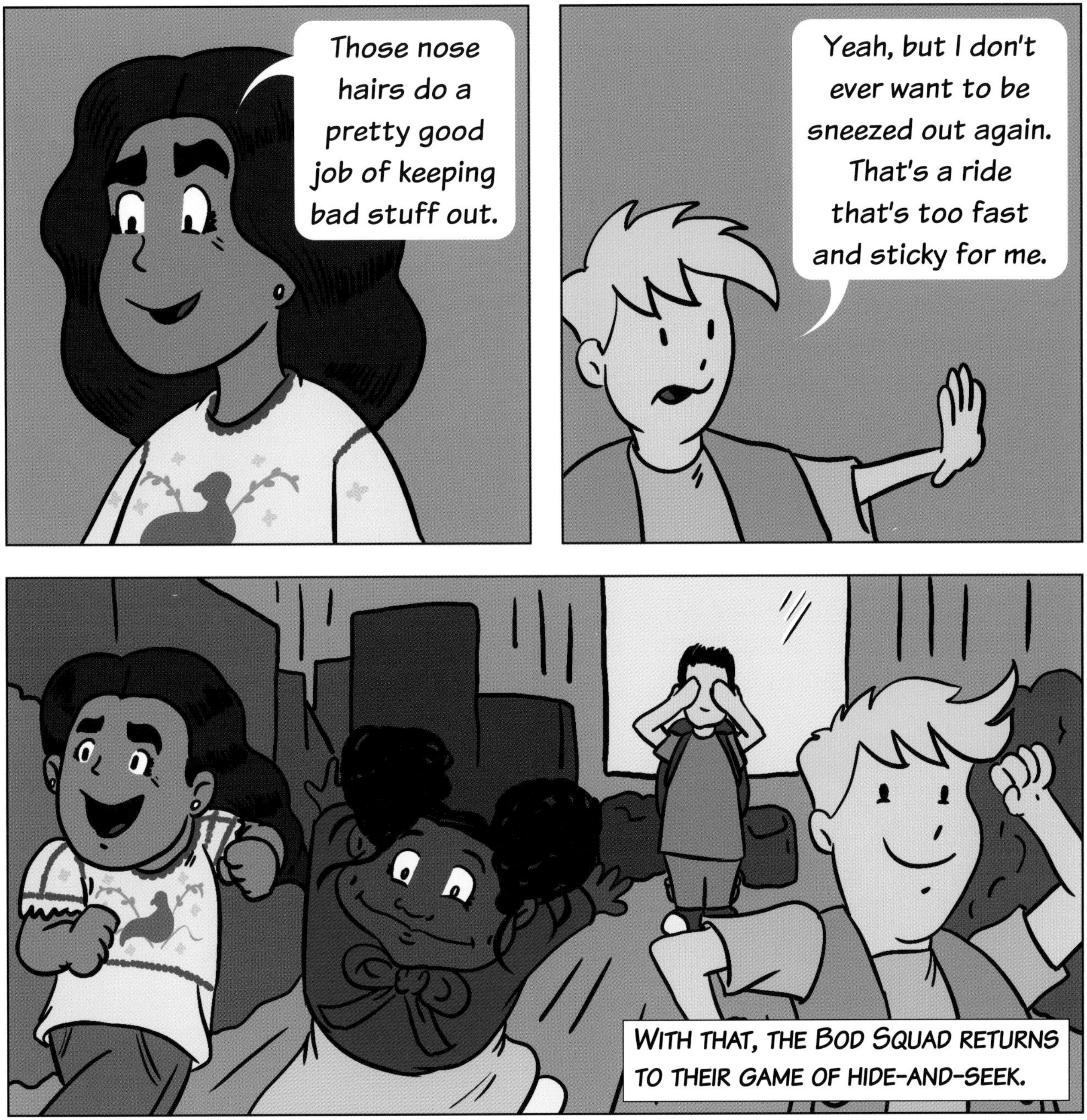
Those nose hairs do a pretty good job of keeping bad stuff out.
Yeah, but I don't ever want to be sneezed out again. That's a ride that's too fast and sticky for me.
WITH THAT, THE BOD SQUAD RETURNS TO THEIR GAME OF HIDE-AND-SEEK.

THE RESPIRATORY SYSTEM

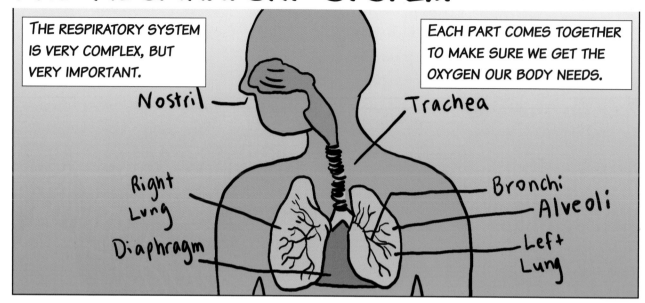

THE RESPIRATORY SYSTEM IS VERY COMPLEX, BUT VERY IMPORTANT.

EACH PART COMES TOGETHER TO MAKE SURE WE GET THE OXYGEN OUR BODY NEEDS.

Nostril

Trachea

Right Lung

Diaphragm

Bronchi

Alveoli

Left Lung

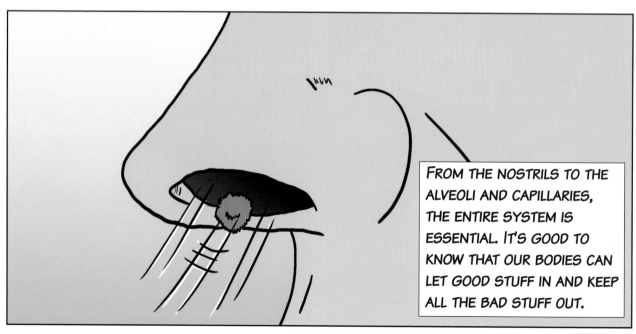

FROM THE NOSTRILS TO THE ALVEOLI AND CAPILLARIES, THE ENTIRE SYSTEM IS ESSENTIAL. IT'S GOOD TO KNOW THAT OUR BODIES CAN LET GOOD STUFF IN AND KEEP ALL THE BAD STUFF OUT.

Glossary

bronchi: tunnels that move air to the lungs

capillaries: the tiniest tubes that move blood throughout the body

carbon dioxide: a gas that comes out of humans and animals when they breathe out

filter: to remove small, unwanted pieces from the air or a liquid

germs: very small living things that can cause illness

molecules: the smallest pieces that a substance can be broken down into

nostril: one of the two holes in the nose

organs: major internal body parts

particles: very small pieces

respiratory: having to do with breathing and processing air

Further Reading

Hunt, Santana. *The Oxygen Cycle.* New York: Gareth Stevens Publishing, 2020. Read about the human body's role in the oxygen cycle.

Lawton, Cassie M. *The Human Respiratory System.* New York: Cavendish Square Publishing, 2021. Learn more about the form and function of the respiratory system.

Mason, Paul. *Your Breathtaking Lungs and Rocking Respiratory System: Find Out How Your Body Works!* New York: Crabtree Publishing Company, 2016. Have all your questions about the respiratory system answered by this colorful book.

GameUp: Guts and Bolts (https://www.brainpop.com/games/gutsandbolts/) Build a robot to learn more about the flow of oxygen in the body.

KidsHealth: Your Lungs and Respiratory System (https://kidshealth.org/en/kids/lungs.html) Read more about how the respiratory system works and how to keep it healthy.

Visible Body: Lower Respiratory System (https://www.visiblebody.com/learn/respiratory/lower-respiratory-system) View a 3-D model of the lower respiratory system, which includes the lungs.

ABOUT THE AUTHOR

Alexander Lowe is a writer who splits his time between Los Angeles and Chicago. He has written children's books about sports, technology, science, and media. He has also done extensive work as a sportswriter and film critic. He loves reading books of any and all kinds.

ABOUT THE ILLUSTRATOR

Sebastian Kadlecik is a screenwriter, actor, and comic book maker. He is best known as the creator of the epic action saga *Penguins vs. Possums*, about a secret, interspecies war for dominion over the earth, and the Eisner-nominated *Quince*, about a young Latina who gets superpowers at her quinceañera.